SpringerBriefs in Neuroscience

For further volumes:
http://www.springer.com/series/8878

Amita Pandey · Girdhar K. Pandey

The UNC-53-mediated Interactome

Analysis of its Role in the Generation of the *C. elegans* Connectome

 Springer

Amita Pandey
Girdhar K. Pandey
Department of Plant Molecular Biology
University of Delhi South Campus
Dhaula Kuan
New Delhi
India

ISSN 2191-558X   ISSN 2191-5598   (electronic)
ISBN 978-3-319-07826-7   ISBN 978-3-319-07827-4   (eBook)
DOI 10.1007/978-3-319-07827-4
Springer Cham Heidelberg New York Dordrecht London

Library of Congress Control Number: 2014940757

© The Author(s) 2014
This work is subject to copyright. All rights are reserved by the Publisher, whether the whole or part of the material is concerned, specifically the rights of translation, reprinting, reuse of illustrations, recitation, broadcasting, reproduction on microfilms or in any other physical way, and transmission or information storage and retrieval, electronic adaptation, computer software, or by similar or dissimilar methodology now known or hereafter developed. Exempted from this legal reservation are brief excerpts in connection with reviews or scholarly analysis or material supplied specifically for the purpose of being entered and executed on a computer system, for exclusive use by the purchaser of the work. Duplication of this publication or parts thereof is permitted only under the provisions of the Copyright Law of the Publisher's location, in its current version, and permission for use must always be obtained from Springer. Permissions for use may be obtained through RightsLink at the Copyright Clearance Center. Violations are liable to prosecution under the respective Copyright Law.
The use of general descriptive names, registered names, trademarks, service marks, etc. in this publication does not imply, even in the absence of a specific statement, that such names are exempt from the relevant protective laws and regulations and therefore free for general use.
While the advice and information in this book are believed to be true and accurate at the date of publication, neither the authors nor the editors nor the publisher can accept any legal responsibility for any errors or omissions that may be made. The publisher makes no warranty, express or implied, with respect to the material contained herein.

Printed on acid-free paper

Springer is part of Springer Science+Business Media (www.springer.com)

Contents

Chapter 1
Introduction

Abstract Metazoan nervous system development involves cell migration and axon outgrowth, where neuronal cell bodies and their axons migrate long distances to reach their final destination to generate the final structure of the nervous system. Migrating cells and axons possess a highly specialized structure at the leading edge called the growth cone composed of three distinct zones, namely the P zone, the T zone, and the C zone. The growth cone perceives the attractive and the repulsive signals in its environment via the receptors expressed on its surface in the P zone, transduce the signal via signal transduction proteins to reorganize the actin cytoskeleton and extend filopodia toward the attractive cues, determining the final movement and direction of the growth cone. Studies conducted in model organism, *Caenorhabditis elegans*, have provided insight into the molecular mechanisms of cell migration and axon outgrowth.

Keywords Nervous system · *C. elegans* · Growth cone · Cell migration · Axon outgrowth · Guidance cues

During metazoan development, many cells undergo cell migrations, including the cells of the gonads, kidney, and the immune and the nervous systems. Nervous system is made up of specialized cells called neurons that have acquired the ability to transmit electrochemical signals in an intricate network of nerve cells, interconnected by synapse. The events leading to the establishment of functional neuronal networks follow a number of key steps, including asymmetric cell division of neuronal precursors or neuroblasts, neuronal cell migration, establishment of polarity, axon outgrowth, and fasciculation. Failure to follow any of the above steps results in a dysfunctional nervous system (NS). Many NS-related disorders have been liked to aberrant proliferation, migration, axon outgrowth, and synaptogenesis including disorders like attention-deficit/hyperactivity disorder (ADHD) (Rivero et al. 2013; Blockus and Chétodal 2014), autism (Anitha et al. 2008; Betancur et al. 2009), and Alzheimer's disease (Zhang and Cai 2010). In order to procure understanding of various developmental

A. Pandey and G. K. Pandey, *The UNC-53-mediated Interactome*,
SpringerBriefs in Neuroscience, DOI: 10.1007/978-3-319-07827-4_1,
© The Author(s) 2014

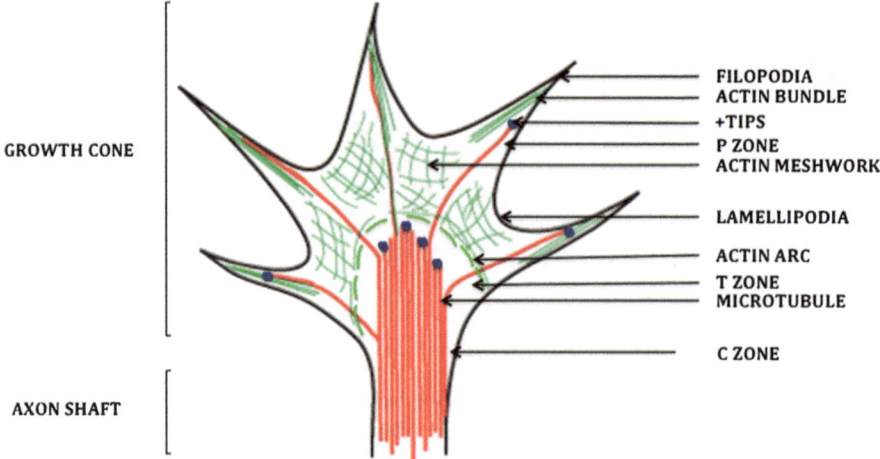

Fig. 1.1 The growth cone morphology: The leading edge of a migrating cell and axon is called the growth cone (*GC*). A GC steers the cells or axons perceiving attractive and repulsive guidance cues in the surrounding environment, responding to them by signaling via plethora of signal transduction proteins to form the final connectivity of the nervous system. A growth cone is divided into three zones, namely peripheral zone (*P zone*), composed of filopodia and lamellipodia, and filopodia are composed of filamentous (F-actin) bundle and microtubule (*MT*); the transition zone (*T zone*), composed of F-actin meshwork and actin arcs; and the central zone (*C zone*), composed of MT bundles. The axon shaft is primarily composed of stable MT bundles. MTs growing ends are associated with plus-end-tracking proteins (+TIPs)

disorders, an understanding of how neural connectivity is established will be important for efforts to improve treatment for nerve connections damaged by injury or neurodegenerative diseases (Yaron and Zheng 2007; Harel and Strittmatter 2006).

One of the most fundamental questions is how cells and axons migrate during NS development? Studies conducted in both invertebrates and vertebrates have revealed that neuronal cell migration and axon outgrowth occur by a highly specialized and dynamic structure, located at the leading edge of migrating cells and axons, called the growth cone (GC), first described by Cajal in 1890, from sections of embryonic spinal cord stained with silver chromate (Cajal 1890). The GC is divided into three different regions (Bridgman and Dailey 1989; Forscher and Smith 1988; Smith 1988), namely the peripheral (P) zone, composed of lamellipodia and filopodia; the transitional (T) zone, a band of the GC at the interface between the P zone and the central (C) zone; and the C zone, composed of thicker regions invested by organelles and vesicles of varying sizes (Fig. 1.1). Filopodia, also called microspikes, are narrow fingerlike extensions, composed of polarized bundled array of filamentous actin (F-actin) and microtubule (MT), extending into the T zone of GC, whereas, lamellipodia are flattened, veil-like extensions at the periphery of the GC, composed of meshwork of F-actin in the P zone (Fig. 1.1) (Bridgman and Dailey 1989; Forscher and Smith 1988; Lewis and Bridgman 1992; Smith 1988). F-actin can adopt dynamic and stable structures in the GC including intrapodia (which originate from T zone and extend into the P zone) (Dent and Kalil 2001; Katoh et al. 1999; Rochlin et al. 1999), arc-like structure (Fig. 1.1) (Schaefer et al. 2002), puncta (central region of the GC and axon shaft), and a thin subplasmalemmal cortical meshwork (axon shaft) (Letourneau 1983; Schnapp and Reese 1982).

Biochemical studies conducted in *Xenopus* retinotectal system (Chien et al. 1993) and *Drosophila* peripheral NS (Kaufmann et al. 1998) supported the requirement of cytoskeleton for directional movement of GC, and cytochalasin-induced F-actin depolymerization resulted in continued extension but defective guidance, whereas depolymerization of microtubules (MTs) by colchicine did not immediately affect filopodia but eventually resulted in retraction of axon (Yamada et al. 1971). Based on these observations, it was concluded that F-actin maintains GC shape and guidance, whereas MTs are essential for axon structure and elongation. Subsequently, biochemical analysis pointed toward an interaction between F-actin and MT, and depolymerization of MTs by colchicine caused new GC-like protrusions along the axon shaft (Bray et al. 1978), suggesting that MT inhibits actin polymerization in the axon shaft. Later, a direct correlation was established between filopodial extensions and GC advancement by quantitative analysis of filopodia from dorsal root ganglion neurons (Bray and Chapman 1985).

Once a correlation was established between the GC movement and coordination of F-actin and MTs, the dynamic cytoskeletal polymers, promoting shape change and locomotion, the next step was to identify the molecules, which served as a link between GC movement and guidance to the cytoskeleton remodeling. In past two decades, studies conducted in model organisms such as *Caenorhabditis elegans* and *Drosophila* led to identification of molecules, regulating the precise directional movement of GC, and proposed that GC movement to its final target requires an orchestrated sequence of steps. First, the GC and surrounding tissue are specified to express a specific combination of receptors and cues, respectively. Second step is expression of receptor complexes on the GC and secretion of attractive and repulsive cues in the extracellular matrix. Third step is perception of signal by the receptors, and signal transduction via cytoplasmic signaling molecules. And the final step involves cytoskeleton remodeling to allow GC to move toward attractive cues and move away from repulsive cues (Fig. 1.2). In vivo, the GC morphology is determined by perception of both attractive and repulsive guidance cues from its surrounding environment.

Over the past two decades, four main guidance cues and their receptors, involved in GC migrations along the dorsal–ventral (DV) axis and anterior–posterior (AP) axis in mammals, flies and nematodes, have been identified. UNC-6, the first guidance cue to be discovered, was identified in *C. elegans* uncoordinated (Unc) mutants with defects in commissural axon outgrowth along the dorsal–ventral axis of the worm (Hedgecock et al. 1990). Later, UNC-6 homologue, Netrin, was identified in mammals and *Drosophila* (Serafini et al. 1994; Harris et al. 1996). Later, studies revealed that UNC-6/Netrin binds to its receptors UNC-40/DCC/FRA and UNC-5/UNC-5 and attracts or repels GCs, respectively (Leung-Hagesteijn et al. 1992; Wadsworth et al. 1996; Chan et al. 1996; Kolodziej et al. 1996). Other highly conserved guidance molecules are the SLT-1/Slit and its repulsive receptor SAX-3/Robo (Kidd et al. 1999;

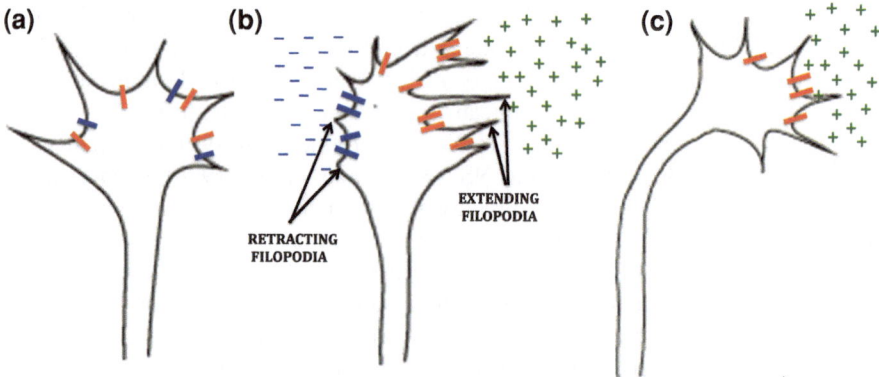

Fig. 1.2 Diagrammatic representation of growth cone at the leading edge of an axon responding to the attractive and repulsive cues: Growth cone is a specialized structure at the leading edge of migrating cells and axons. **a** A growth cone expressing receptors for both chemoattractant and chemorepellent in the P zone. **b** Growth cone perceives the extracellular attractive and repulsive cues via receptor complexes and moves in the direction of attractive cue by extending filopodia and is repelled away from repulsive cue by retracting filopodia. **c** Growth cone advancement in the direction of attractive cue by expressing receptors for attractive cue on its surface

Hao et al. 2001), and Wnts and their receptor Frizzled (Fz) and Drl/Ryk (Silhankova and Korswagen 2007) that mediate GC movement and guidance. After the signal perception by receptors at the GC surface, it is transduced via cytoplasmic signal-transducing proteins to initiate actin cytoskeleton remodeling. The signaling molecules associated in this process include small GTPases, phosphoinositide 3-kinase (PI3K), and calcium, acting downstream of receptors, and these in turn activate effector proteins, which transduce signal to actin modulatory proteins. Scaffolding proteins or adaptor proteins are associated in the formation of signal transduction complexes or modules at the membrane or in the cytoplasm. These signaling modules or interactomes promote reorganization of actin and microtubule cytoskeleton, resulting in polarity establishment and initiation of directional GC movement.

Choice of the system to study the process of axon outgrowth and cell migration depends upon the number of migrating cells and axons in an organism. The human and mouse NS are highly complex due to numerous cells undergoing migration during development; therefore, relatively less complex organisms have been used for studying this process including *Drosophila* (fruit fly) and *C. elegans* (nematode). Out of these two, the simpler body plan of *C. elegans* makes it an excellent model organism to study cell migration and axon outgrowth. The features, which make *C. elegans* an ideal organism to study cell migration and axon outgrowth, are its simple anatomy and it is eutelic, that is all individuals have 952 cells of which 302 are neurons. It has invariant pattern of cell division from the zygote to adult, due to its small size (1 mm) and being transparent, the position of processes and putative synaptic connections of all the cells are well documented from serial section electron micrographs (White et al. 1986). Moreover, *C. elegans* is a multicellular

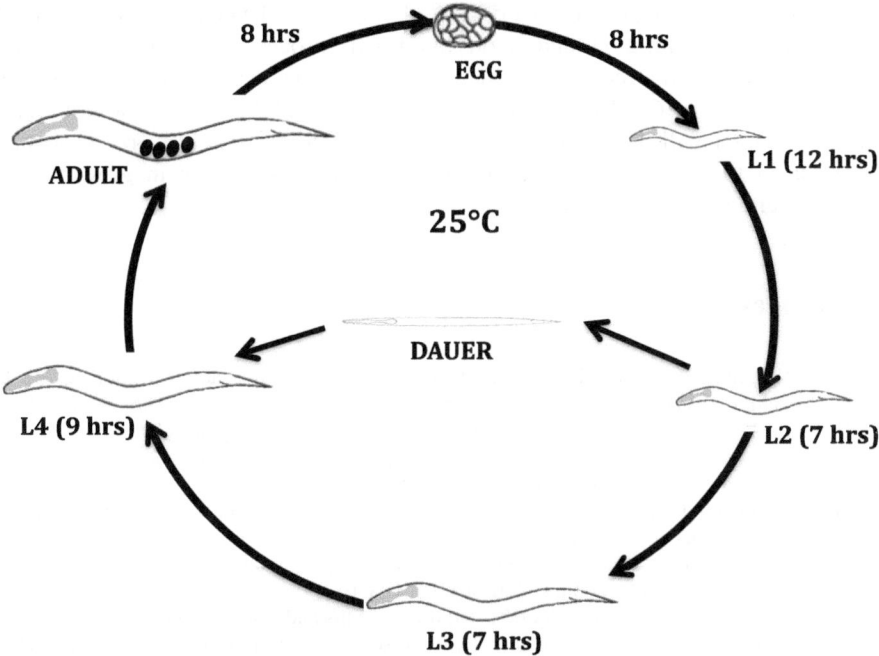

Fig. 1.3 Life cycle of *C. elegans*: *C. elegans* hermaphrodite life cycle is dependent on temperature, and it completes life cycle in 2.5 days at 25 °C. It has different larval stages (L1, L2, L3, and L4), and adult worms can produce up to 300 progeny over their life span. Dauer stage is formed when worms are exposed to stress conditions like starvation and high temperature (over 27 °C)

organism and follows the same complex developmental processes like humans such as embryogenesis and morphogenesis; therefore, the information obtained from *C. elegans* can be directly applied to more complex organisms including human. For example, during development, relatively fewer cells migrate in *C. elegans* as opposed to *Drosophila* or mouse (Blelloch et al. 1999; Forrester et al. 1998; Montell 1998). Migrations and axon outgrowth are invariant from one animal to another, so any kind of divergence from normal pattern can be easily identified. Thirty-five percent of *C. elegans* genes have human homologues, and human genes can functionally replace their worm homologue when introduced into the worm. For example, many of the cues and receptors regulating migration and axon outgrowth in *C. elegans* have human homologues. *C. elegans* genome size is relatively small (9.7×10^7 base pairs or 97 Mb), when compared to the human genome, which is estimated to consist of 3 billion base pairs (3×10^9 bp or 3,000 Mb), and a complete annotated nematode genome sequence is available. Simple routine molecular biology techniques can be used to study gene function including reducing gene function by RNA-mediated interference (RNAi). Lastly, *C. elegans* is very simple to maintain in the laboratory with a life span of 2–3 weeks, and development to adult stage takes 2.5 days at 23 °C (Fig. 1.3).

The subsequent chapters in this book will discuss guidance molecules involved in growth cone migration and axon outgrowth process, and the interactome involved in linking the receptors at the membrane to the actin cytoskeleton during NS development in *C. elegans* and their vertebrate and *Drosophila* homologues. The last chapter will discuss the cytoplasmic scaffolding protein UNC-53, which functions in the longitudinal migrations in worms and its role in growth cone migration and axon outgrowth process.

References

Anitha A, Nakamura K, Yamada K, Suda S, Thanseem I, Tsujii M, Iwayama Y, Hattori E, Toyota T, Miyachi T, Iwata Y, Suzuki K, Matsuzaki H, Kawai M, Sekine Y, Tsuchiya K, Sugihara G, Ouchi Y, Sugiyama T, Koizumi K, Higashida H, Takei N, Yoshikawa T, Mori N (2008) Genetic analyses of roundabout (ROBO) axon guidance receptors in autism. Am J Med Genet B Neuropsychiatr Genet. 147B(7):1019–1027

Betancur C, Sakurai T, Buxbaum JD (2009) The emerging role of synaptic cell-adhesion pathways in the pathogenesis of autism spectrum disorders. Trends Neurosci 32(7):402–412

Blelloch R, Newman C, Kimble J (1999) Control of cell migration during *Caenorhabditis elegans* development. Curr Opin Cell Biol 11(5):608–613

Blockus H, Chédotal A (2014) The multifaceted roles of Slits and Robos in cortical circuits: from proliferation to axon guidance and neurological diseases. Curr Opin Neurobiol. 27C:82–88

Bray D, Chapman K (1985) Analysis of microspike movement on the neuronal growth cone. J Neuroscience 5:3204–3213

Bray D, Thomas C, Shaw G (1978) Growth cone formation in cultures of sensory neurons. Proc Natl Acad Sci USA 75:5226–5229

Bridgman PC, Dailey ME (1989) The organization of myosin and actin in rapid frozen nerve growth cones. J Cell Biol 108:95–109

Cajal SRY (1890) À quelle é poque apparaissent les expansions des cellules nerveuses de la moë lle é piniè re du poulet? Anatomo-mischer Anzeiger 21–22:609–639

Chan SS, Zheng H, Su MW, Wild R, Killeen MT, Hedgecock EM, Culotti JG (1996) UNC-40, a *C. elegans* homolog of DCC (deleted in colorectal cancer), is required in motile cells responding to UNC-6/Netrin cues. Cell 87:187–195

Chien CB, Rosenthal DE, Harris WA, Holt CE (1993) Navigational errors made by growth cones without filopodia in the embryonic *Xenopus* brain. Neuron 11:237–251

Dent EW, Kalil K (2001) Axon branching requires interactions between dynamic microtubules and actin filaments. J Neurosci 21:9757–9769

Forrester WC, Perens E, Zallen JA, Garriga G (1998) Identification of *Caenorhabditis elegans* genes required for neuronal differentiation and migration. Genetics 148(1):151–165

Forscher P, Smith SJ (1988) Actions of cytochalasins on the organization of actin filaments and microtubules in a neuronal growth cone. J Cell Biol 107:1505–1516

Harel NY, Strittmatter SM (2006) Can regenerating axons recapitulate developmental guidance during recovery from spinal cord injury? Nat Rev Neurosci. 7(8):603–616

Hao JC, Yu TW, Fujisawa K, Culotti JG, Gengyo-Ando K, Mitani S, Moulder G, Barstead R, Tessier-Lavigne M, Bargmann CI (2001) *C. elegans* Slit acts in midline, dorsal–ventral, and anterior–posterior guidance via the SAX-3/Robo receptor. Neuron 32:25–38

Harris R, Sabatelli LM, Seeger MA (1996) Guidance cues at the *Drosophila* CNS midline: identification of two *Drosophila* Netrin UNC-6 homologs. Neuron 17:217–228

Hedgecock EM, Culotti JG, Hall DH (1990) The *unc-5*, *unc-6*, and *unc-40* genes guide circumferential migrations of pioneer axons and mesodermal cells on the epidermis in *C. elegans*. Neuron 4:61–85

Katoh K, Hammar K, Smith PJ, Oldenbourg R (1999) Arrangement of radial actin bundles in the growth cone of *Aplysia* bag cell neurons shows the immediate past history of filopodial behavior. Proc Natl Acad Sci USA 96:7928–7931

Kaufmann N, Wills ZP, Van Vactor D (1998) *Drosophila* Rac1 controls motor axon guidance. Development 125:453–461

Kidd T, Bland KS, Goodman CS (1999) Slit is the midline repellent for the robo receptor in *Drosophila*. Cell 96(6):785–794

Kolodziej PA, Timpe LC, Mitchell KJ, Fried SR, Goodman CS, Jan LY, Jan YN (1996) Frazzled encodes a *Drosophila* member of the DCC immunoglobulin subfamily and is required for CNS and motor axon guidance. Cell 87:197–204

Letourneau PC (1983) Differences in the organization of actin in the growth cones compared with the neurites of cultured neurons from chick embryos. J Cell Biol 97:963–973

Leung-Hagesteijn C, Spence AM, Stern BD, Zhou Y, Su MW, Hedgecock EM, Culotti JG (1992) UNC-5, a transmembrane protein with immunoglobulin and thrombospondin type 1 domains, guides cell and pioneer axon migrations in *C. elegans*. Cell 71(2):289–299

Lewis AK and Bridgman PC (1992) Nerve growth cone lamellipodia contain two populations of actin filaments that differ in organization and polarity. J Cell Biol 119:1219–1243

Montell C (1998) TRP trapped in fly signaling web. Curr Opin Neurobiol 8(3):389–397

Rivero O, Sick S, Popp S, Schmitt A, Franke B, Lesch KP (2013) Impact of ADHD-susceptibility gene CDH13 on development and function of brain networks. Eur Neuropsychopharmacol 23(6):492–507

Rochlin MW, Dailey ME, Bridgman PC (1999) Polymerizing microtubules activate site-directed F-actin assembly in nerve growth cones. Mol Biol Cell 10:2309–2327

Schaefer AW, Kabir N, Forscher P (2002) Filopodia and actin arcs guide the assembly and transport of two populations of microtubules with unique dynamic parameters in neuronal growth cones. J Cell Biol 158:139–152

Schnapp BJ, Reese TS (1982) Cytoplasmic structure in rapid-frozen axons. J Cell Biol 94:667–669

Serafini T, Kennedy TE, Galko MJ, Mirzayan C, Jessell TM, Tessier-Lavigne M (1994) The Netrins define a family of axon outgrowth-promoting proteins homologous to *C. elegans* UNC-6. Cell 78:409–424

Silhankova M, Korswagen HC (2007) Migration of neuronal cells along the anterior–posterior body axis of *C. elegans*: Wnts are in control. Curr Opin Genet Dev 17:320–325

Smith SJ (1988) Neuronal cytomechanics: the actin-based motility of growth cones. Science 242:708–715

Wadsworth WG, Bhatt H, Hedgecock EM (1996) Neuroglia and pioneer neurons express UNC-6 to provide global and local netrin cues for guiding migrations in *C. elegans*. Neuron 16:35–46

White JG, Southgate E, Thomson JN, Brenner S (1986) The structure of the nervous system of the nematode *Caenorhabditis elegans*. Phil Trans R Soc B 314:1–340

Yamada KM, Spooner BS, Wessells NK (1971) Ultrastructure and function of growth cones and axons of cultured nerve cells. J Cell Biol 49:614–635

Yaron A, Zheng B (2007) Navigating their way to the clinic: emerging roles for axon guidance molecules in neurological disorders and injury. Dev Neurobiol 67(9):1216–1231

Zhang J, Cai H (2010) Netrin-1 prevents ischemia/reperfusion-induced myocardial infarction via a DCC/ERK1/2/eNOSs1177/NO/DCC feed-forward mechanism. J Mol Cell Cardiol 48(6):1060–1070

Chapter 2
Guidance Molecules Required for Growth Cone Migration of Cells and Axons

Abstract In metazoans, neuronal cell bodies and their axons migrate long distances along both the dorsal–ventral and the anterior–posterior axis during different phases of development; additionally, they may also cross the midline in bilaterally symmetric organisms. The leading edge of a migrating cell or an axon is a highly specialized structure, the growth cone (GC). The GC steers through an increasingly complex environment by perceiving attractive and repulsive cues via their cognate receptors expressed on the GC surface. Interestingly, studies in the past two decades revealed that the axon guidance molecules are highly conserved across eukaryotes including UNC-6/Netrins, acting as a bifunctional cue, attracting and repelling GCs via UNC-40/DCC and UNC-5/UNC-5 receptors, respectively, SLT-1/Slit repelling GCs via its receptor SAX-3/Robo, and Wnts acting through their Frizzled (Fz) and Ryk receptors. Besides these universally conserved guidance molecules, there are molecules specific to nematodes steering the GCs during nervous system development.

Keywords Guidance cues · cell migration · Axon outgrowth · UNC-6 · Netrin · SLIT · Robo · Wnts · UNC-53

Studies in mammals, flies, and nematodes led to identification of molecules required for guidance and extension of growth cone (GC) during nervous system development. The following section discusses the major cues and their cognate receptors required for circumferential or dorsal–ventral (DV), longitudinal or anterior–posterior (AP) GC guidance and midline guidance in bilaterally symmetric organisms.

A. Pandey and G. K. Pandey, *The UNC-53-mediated Interactome*,
SpringerBriefs in Neuroscience, DOI: 10.1007/978-3-319-07827-4_2,
© The Author(s) 2014

2.1 Guidance Cues and Their Receptors in Dorsal–Ventral Guidance

2.1.1 UNC-6/Netrin and UNC-40/DCC/FRA

A genetic screen for mutations affecting circumferential migrations in *Caenorhabditis elegans* led to identification of three genes *unc-6*, *unc-5*, and *unc-40*, where *unc-5* and *unc-40* gene products affect ventral and dorsal migrations of the GCs of pioneer axons and mesodermal cell along the epidermis in an *unc-6*-dependent manner causing the uncoordinated (Unc) phenotype (Hedgecock et al. 1987, 1990). UNC-6 is ventrally localized laminin-related extracellular matrix protein, expressed in twelve types of neuroglia and pioneer neurons (Hedgecock et al. 1990; Chan et al. 1996), affecting both dorsal and ventral migrations (Wadsworth et al. 1996). UNC-6 serves as a bifunctional guidance cue in both vertebrates and invertebrates, acting both as a chemorepellent and as a chemoattractant (Hedgecock et al. 1990; Tessier-Lavigne and Goodman 1996). The vertebrate, UNC-6 homologues, Netrin-1 (*Ntn-1*) and Netrin-2 (*Ntn-2*), expressed at the ventral midline of chick embryonic spinal cord (Serafini et al. 1994), guide commissural axons toward the ventral midline and telencephalon, acting as attractive cue (Métin et al. 1997) and repulsive cue in guiding trochlear motor axons dorsally away from the ventral floor plate and GABAergic interneurons in the cerebellar cortex (Colamarino and Tessier-Lavigne 1995; Guijarro et al. 2006). Netrin-1 acts as guidance cue in several regions of the central nervous system (CNS), including spinal commissural, hippocampal, retinal, thalamic, and dopaminergic axon guidance (Kennedy et al. 1994; Colamarino and Tessier-Lavigne 1995; Barallobre et al. 2000; Braisted et al. 2000; Shewan et al. 2002; Lin et al. 2005). Similarly, the *Drosophila* homologues, Netrin-A and Netrin-B, expressed at the ventral midline, function as an attractive cue during commissure formation and repulsive cue, repelling the motor axon outgrowth to their targets (Harris et al. 1996; Mitchell et al. 1996). Mutations in Netrin caused partially missing or thinner commissures, a phenotype that is rescued by expression of Netrin from the ventral midline.

UNC-6/Netrin achieves its chemoattractant ventral guidance function through its receptor UNC-40 of *C. elegans*, Neogenin/DCC (deleted in colorectal cancer) of vertebrates, and FRA (Frazzled) of *Drosophila*, expressed on motile cells and pioneer neurons (Hedgecock et al. 1990; Fearon et al. 1990; Keino-Masu et al. 1996; Kolodziej et al. 1996; Chan et al. 1996). *DCC* was originally identified as a tumor suppressor gene, deleted at high frequency in colorectal cancers (Fearon et al. 1990), expressed on the axons of the dorsal spinal commissural neurons as they extend across the floor plate and to the ventral funiculus. Biochemical assays using antibodies against DCC supported the function of DCC as a receptor for Netrin-1, where Netrin-1-dependent commissural axon outgrowth was stopped in in vitro assays (Keino-Masu et al. 1996). Similarly, *frazzled* (*fra*), *Drosophila* member of *DCC* gene family, was also found to be expressed on the axons in the embryonic CNS and on motor axons in the periphery, null mutations in *fra*

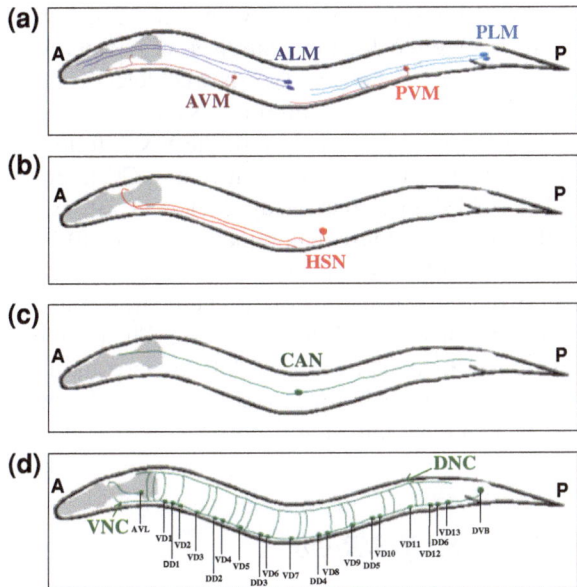

Fig. 2.1 Illustrative of neuronal cell bodies and their axons during *C. elegans* development: **a** The six mechanosensory neurons including two anterior lateral microtubules (*ALMs*), located in the midbody region, and two posterior lateral microtubules (*PLMs*), located in the tail of the worm. ALM extends anterior processes ending in the first pharynx bulb, and PLMs are bipolar with a shorter posterior process extended into the tail and an anterior process terminating near the *ALM* cell body. Anterior ventral microtubule and posterior ventral microtubule (*AVM* and *PVM*) extend a ventral process to the ventral nerve cord (*VNC*), followed by anterior outgrowth toward the head of the worm. **b** Hermaphrodite-specific neurons (*HSNs*), a pair of bilaterally symmetric neuron, with the cell body located in the midbody region of the worm, extends a ventral process to the VNC, which later extends anteriorly to the nerve ring. **c** Canal-associated neurons (*CANs*), with their cell bodies located in the midbody region of the worm extending processes in both anterior and posterior directions. **d** The six DD and thirteen VD neurons send a commissural axonal process during embryogenesis (*DDs*) or larval development (*VDs*), and commissures extend from the VNC toward the dorsal nerve cord (*DNC*)

disrupted axon guidance, and the phenotypes resembled those observed in embryos lacking the *Drosophila* Netrin genes (Harris et al. 1996; Mitchell et al. 1996). Additionally, expression of FRA in the neurons rescued the defects in motor axon guidance, strongly supporting it to be a receptor for Netrin (Kolodziej et al. 1996).

In *C. elegans*, *unc-40* gene function is required for GC migrations of several cells including AVM and PVM (anterior and posterior ventral microtubule neuron) mechanosensory or touch receptor neurons that sense gentle body touch (Fig. 2.1a) (Chalfie and Sulston 1981; Chalfie et al. 1985). In wild-type animals, AVM cell body is located laterally in the posterior left side of the worm body, whereas PVM cell body is located laterally in anterior right side of the worm body, and both the neurons send out a ventral process to the ventral nerve cord (VNC), followed by anterior longitudinal extension along the VNC. Phenotypic analysis of *mec-4:GFP* transgenic worms with *unc-40* mutation revealed that function of UNC-40 is required for the ventral process outgrowth (Chalfie et al. 1994; Chan et al. 1996). *unc-40* mutants also showed defects in the ventral process outgrowth of hermaphrodite-specific neuron (HSN), a pair of bilaterally symmetric neurons extending a ventral process toward the VNC, followed by an anterior process toward the nerve ring (Fig. 2.1b) and in ventral GC migrations of the phasmid (PHA and PHB) and PDE neurons (Hedgecock et al. 1990).

Several lines of evidence suggest requirement of UNC-40 in both ventral and dorsal GC migrations. SDQR a sister neuron of AVM, born in the first larval stage, migrates dorsally to the lateral margin of the body wall muscles, and its axon grows dorsally and anteriorly toward the nerve ring. Phenotypic analysis of *unc-119::GFP* transgenic worms (Chalfie et al. 1994; Maduro and Pilgrim 1995) with *unc-40* mutation revealed defects in the dorsal axon guidance but less severe than the *unc-6* and *unc-5* mutations (Kim et al. 1999).

2.1.2 SLT-1/Slit and SAX-3/Robo

Another ventral guidance pathway conserved across species is the SLT-1/Slit (Shiga-like toxin) and its cognate receptor SAX-3/Robo (Roundabout). Initially, *slit* gene was identified in *Drosophila* embryo as a larval cuticle-patterning gene (Nüsslein-Volhard et al. 1984), and later, it was shown that null mutations in *slit* gene resulted in disruption of *Drosophila* embryonic CNS (Rothberg et al. 1988). Antibody staining and in situ hybridization revealed *slit* expression at midline glial cells, and reduction in *slit* expression resulted in disruption of developing midline cells and commissural axon pathways (Rothberg et al. 1990). Roundabout (*robo*), receptor for Slit, was identified along with *slit* and *comm* in a large-scale mutant screen for the genes that control midline axon guidance in *Drosophila* CNS (Seeger et al. 1993). Axons expressing high Robo levels from the outset never crossed the midline, and axons that crossed the midline once, expressed Robo after crossing, suggesting that Robo acts as a repulsive receptor (Kidd et al. 1998a). Phenotypic analysis of transhet-erozygotes and transgenic for *slit* and *robo* confirmed their role in a common genetic pathway, where Slit acts as a ligand to Robo in midline axon guidance in *Drosophila* embryos (Kidd et al 1998b; Kidd et al. 1999). Genetic interaction between Slit and Robo was supported by biochemical studies confirming their physical interaction (Brose et al. 1999). Later, a single *slit* gene was identified in *C. elegans* (*slt-1*) and three *slit* genes in humans (*slit-1–slit-3*) (Itoh et al. 1998; Brose et al. 1999).

In *C. elegans*, SLT-1 protein is expressed in larval dorsal muscles and signals through its receptor SAX-3/Robo to promote ventral guidance of GCs (Hao et al. 2001). SAX-3/Robo is required for ventral process guidance of HSN motor neurons (Fig. 2.1b), evidenced by phenotypic analysis of *sax-3* mutants, where HSN fails to extend a ventral process (Zallen et al. 1998). Besides, genetic analysis supported the requirement of SAX-3/Robo for ventral process guidance of PVM and AVM touch neurons (Fig. 2.1a), in SLT-1/Slit-dependent manner (Hao et al. 2001).

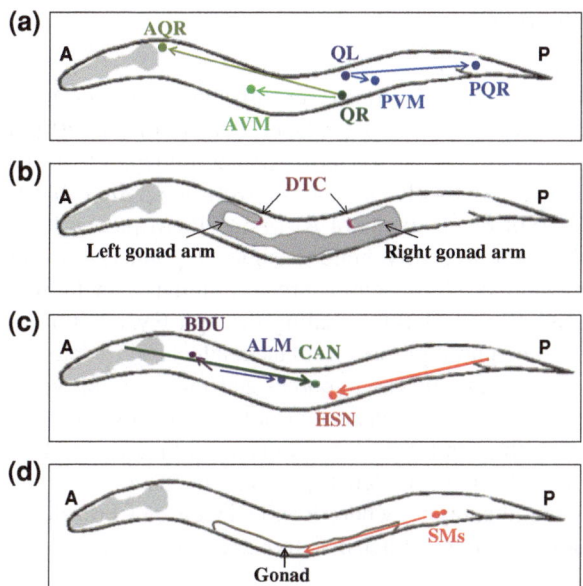

Fig. 2.2 Diagrammatic representation of neuronal and non-neuronal cell migration during *C. elegans* development: **a** Q neuroblasts include *QL* (*left*) and *QR* (*right*): QR migrates anteriorly and divides into *AQR* and *AVM*, and *QL* migrates posteriorly and divides into *PVM* and *PQR*. **b** Distal tip cells (*DTC*), a pair of cells at the tip of each gonad arm, migrate during larval stage away from gonad premordium on the ventral side and turn dorsally and finally migrate toward the center on dorsal side forming a U-shaped gonad. **c** Anterior lateral microtubule (*ALM*) and its sister cell BDU migrate during embryogenesis. ALM undergoes posterior migration to the midbody region of the animal, completing its migration by L1 stage, and BDU migrates anteriorly. **d** The two sex myoblasts (*SMs*) are born in the posterior body region in L1 stage and migrate anteriorly to the midbody region to flank the gonads in the L2 stage

In *C. elegans*, SAX-3/Robo is also required to maintain the sex myoblasts (SMs) along the ventral muscle quadrants (Fig. 2.2d) (Branda and Stern 2000). SMs are bilaterally symmetric pair of cells born during early larval development in the ventral midposterior body region, and after birth, they migrate anteriorly on ventral muscle quadrant to flank the precise center of the gonads, where they divide and differentiate into the muscles required for egg laying (Sulston and Horvitz 1977). Three pathways regulate the anterior migration of SMs including gonad-dependent attraction (GDA), gonad-independent mechanism (GIM), and gonad-dependent repulsion (GDR) (Thomas et al. 1990; Chen et al. 1997; Branda and Stern 2000). Phenotypic analysis of mutants confirmed the function of *egl-17* and *egl-15* in GDA (Stern and Horvitz 1991; DeVore et al. 1995); *unc-53*, *unc-71*, and *unc-73* in GIM (Chen et al. 1997); and *unc-14*, *unc-33*, *unc-44*, and *unc-51* in GDR (Branda and Stern 2000). Moreover, it was found that SAX-3/Robo acts redundantly with GDA to position the SMs in the ventral position, supported by phenotypic analysis of *sax-3*; *egl-17* mutants and *sax-3* mutation with gonad ablation experiments, with more dorsally displaced SMs, were observed (Branda and Stern 2000).

2.1.3 SLT-1/EVA-1

A second SLT-1 receptor has been recently identified in *C. elegans*, called enhancer of ventral axon (EVA-1) guidance defects of *unc-40* mutants, required for ventral process guidance of pioneer AVM neuron (Fig. 2.1a). EVA-1 functions as a co-receptor with SAX-3/Robo. Earlier work supported that SAX-3/Robo in response to SLT-1/Slit repels axons of AVM and PVM pioneering neurons toward VNC (Hao et al. 2001). Recently, genetic evidence pointed toward the presence of two parallel pathways in AVM ventral process guidance including UNC-6/Netrin signaling via UNC-40 and SLT-1 signaling through EVA-1. Based on the genetic evidence where more ventral process guidance defects were detected in *eva-1, slt-1* and then in *sax-3* mutants, it was hypothesized that in wild-type worms, SAX-3 binds to EVA-1, whereas in the absence of either SLT-1 or EVA-1, SAX-3 binds to UNC-40, inhibiting the ventral process guidance (Fujisawa et al. 2007). This proposed inhibition of the UNC-6/Netrin pathway by SAX-3 is consistent with the fly midline crossing phenotype, where Robo silencing of Frazzled/UNC-40 allows GCs to cross midline sources of netrin (Stein and Tessier-Lavigne 2001).

2.1.4 UNC-6/Netrin and UNC-5/UNC-5

unc-5 was identified as one of the three genes affecting circumferential migration of pioneer axons and mesodermal cells on *C. elegans* body wall, specifically affecting dorsal migration (Hedgecock et al. 1990). Sequence analysis revealed UNC-5 as a transmembrane protein, expressed on cell surface of pioneer axons and motile cells, repelled by UNC-6 gradient (Hedgecock et al. 1990; Leung-Hagesteijn et al. 1992; Wadsworth et al. 1996). Three vertebrate homologues of *unc-5*, namely *UNC5h1, UNC5h2,* and *UNC5h3* have been identified in mouse and human (Leonardo et al. 1997), acting as repulsive receptor for Netrin-1, regulating migration of cerebellar neurons and corticospinal tract (CST) axon (Ackerman and Knowles 1998; Finger et al. 2002). A mouse homologue of UNC-5, rostral cerebellar malformation (*rcm*) gene, was identified, and *rcm* mutants exhibited cerebellar and midbrain defects, apparently as a result of abnormal neuronal migration (Ackerman et al. 1997). Similarly, *Drosophila UNC-5* was shown to be required for motor axon guidance, and ectopic expression of UNC-5 in CNS caused midline repulsion of axons (Keleman and Dickson 2001). Interestingly, some of the migrations regulated by UNC-5 required DCC function, evidenced by experiments conducted in *Xenopus* spinal axons, where UNC-5-mediated repulsion was achieved by the formation of receptor complex between DCC and UNC-5, in response to Netrin-1. The interaction between the cytoplasmic domains converted the DCC-mediated attraction to UNC-5-/DCC-mediated repulsion (Hong et al. 1999).

In *C. elegans*, phenotypic analysis of *unc-5* single mutants and *unc-5 unc-6* double mutants revealed that *unc-5* gene function requires *unc-6* in various dorsal migrations including motor neuron axons (Fig. 2.1d), extending dorsal

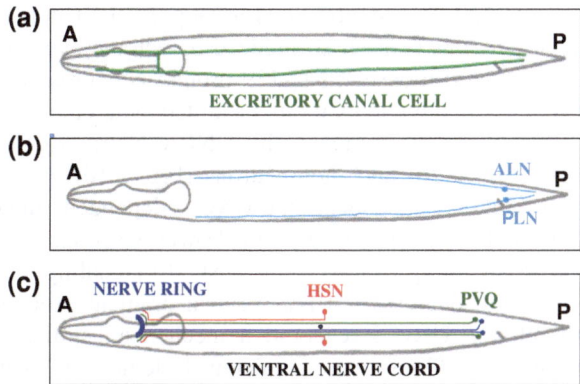

Fig. 2.3 Delineative of neuronal and non-neuronal cell migration and axon outgrowth during *C. elegans* development: **a** The excretory canal is a H-shaped cell, extending out two posterior and two anterior processes with its cell body located in the second pharynx bulb. **b** *ALN* and *PLN* neurons, with their cell bodies located in the tail, extend a long anterior process toward the head and a short posterior process to the tail of the worm. **c** The *C. elegans* ventral nerve cord (*VNC*) is bilaterally asymmetric with the four neurons in the right bundle and forty neurons in the left bundle. The axons in the right bundle cross over the ventral midline to the contralateral side (left bundle), whereas the axons in the left bundle are always ipsilateral and never cross the midline. All the cell bodies send axons to the nerve ring located between the first and second pharynx bulb

commissures from cell bodies located in the VNC (Fig. 2.1d), SDQR, excretory canal cell (EC) (Fig. 2.3a), and distal tip cells (DTCs) (Fig. 2.2b) (Hedgecock et al. 1990; Kim et al. 1999). The role of UNC-5 in dorsal guidance was further supported by ectopic expression of UNC-5 in touch neurons including AVM and PVM (Fig. 2.1a), where the ventral processes were dorsally rerouted (Hamelin et al. 1993), suggesting that UNC-5 is sufficient and necessary for dorsal GC guidance and migration. Analogous to vertebrates where function of UNC-40 is required in guidance mediated by UNC-5, genetic analysis of DTC migration in hypomorphic *unc-5* allele in an *unc-40* wild-type and null background led the authors to propose that UNC-5 and UNC-40 receptors mediate chemorepulsion either independently or together, and this could be due to the formation of various oligomeric receptor complexes (Merz et al. 2001).

2.1.5   UNC-129/TGF-β/BMP and UNC-5/UNC-5

Another gene implicated in migration along the DV axis of the worms is *unc-129*, encoding a TGF-β (transforming growth factor-β)/BMP (bone morphogenetic protein) superfamily molecule, a 407-aa protein (Colavita et al. 1998). In mouse, BMPs are expressed in the dorsal roof plate of developing spinal cord, opposite to Netrin-1, and repel commissural axons toward the ventral floor plate or Netrin-1 source and guide motor neuron away from ventral floor plate, requiring DCC/UNC-40 function (Augsburger et al. 1999; Dillon et al. 2007).

In *C. elegans*, *unc*-129 was identified as a suppressor of dorsal migration of touch receptor neuron axons (Fig. 2.1a) induced by ectopic expression of UNC-5 in these cells, analogous to mouse, UNC-129 protein is expressed in dorsal rows of body wall muscles (Colavita and Culotti 1998; Colavita et al. 1998). UNC-129 affects GC migrations that are regulated by UNC-6/Netrin signaling, including migration of DTC (Fig. 2.2b) and motoneurons (Fig. 2.1d) in a TGF-β serine/threonine receptor kinase-independent but UNC-5-dependent manner (Colavita et al. 1998; Hedgecock et al. 1990; Qin and Powell-Coffman 2004). UNC-129 switches the UNC-5 signaling to UNC-5 + UNC-40 signaling so as to increase the sensitivity of GC to decreasing UNC-6/netrin gradient probably by binding to UNC-5 or mobilizing UNC-5 receptor, a mechanism similar to *Drosophila* UNC-5-mediated short- and long-range repulsion (Keleman and Dickson 2001).

2.1.6 UNC-6/Netrin and ENU-3

C. elegans ENU-3 (enhancer of Unc), a predicted transmembrane protein with a signal peptide and a coiled-coil region (Yee et al. 2011), was identified in a genetic enhancer screen in *unc-5* mutants, as an enhancer of dorsal axon outgrowth and guidance defects of motoneurons (Fig. 2.1d). The motoneuron cell bodies are located in the VNC and they send out a dorsal process toward the dorsal nerve cord (DNC), a process regulated by UNC-5 (Leung-Hagesteijn et al. 1992). *enu-3* mutant worms showed enhanced axon outgrowth defects of the DB4, DB5, and DB6 motor neuron axons in *unc-5* and *unc-6* loss-of-function strains (Yee et al. 2011). Mutations in *enu-3* also enhanced the axon migration defects of the motor neuron axons in a hypomorphic strain with defects in UNC-5-mediated migrations of GC, suggesting that ENU-3 and UNC-5 work together in motor axon guidance in an UNC-6-dependent manner (Yee et al. 2011). Besides, *enu-3* mutation enhanced the defects in ventral process guidance of AVM and PVM touch receptor neurons (Fig. 2.1a) in strains lacking *unc-40* and *slt-1* function. These observations suggest that ENU-3 and UNC-40 function in parallel pathways dependent on UNC-6 (Yee et al. 2014), where ENU-3 either functions as a receptor or modulates the function of an unidentified UNC-6 receptor.

Some other molecules implicated in dorsal–ventral guidance include MADD-4 (muscle arm development defective) (Seetharaman et al. 2011) and MADD-2 (Alexander et al. 2010), both acting in UNC-40 attractive pathway.

2.2 Cues and Receptors in Midline Guidance

In organisms with bilaterally symmetric nervous system including insects and vertebrates, a midline structure establishes a partition between the two mirror image halves. Axons that connect these two halves cross the midline forming axon commissures and are called commissural axons. Commissural axons cross the midline once to leave it on the contralateral side, whereas the ipsilateral axons never cross the midline. Once commissural axons have crossed the midline, they continue to grow longitudinally along the midline never recrossing again.

Studies in Nematodes, flies, rodents, and chick embryos provided insight into the cues and receptors regulating the guidance of commissural and ipsilateral axon GCs, indicating that the midline secretes both attractive and repulsive cues including Netrin/UNC-6 and Slit/SLT-1, in mammals, zebrafish, *Drosophila*, and *C. elegans* (Serafini et al. 1994; Harris et al. 1996; Mitchell et al. 1996; Wadsworth et al. 1996; Battye et al. 1999; Kidd et al. 1999; Yuan et al. 1999; Hao et al. 2001; Yeo et al. 2001). In vertebrates and *Drosophila*, Netrin–DCC/FRA and Slit–Robo function in ventral midline guidance. The midline Netrin/UNC-6 source attracts the commissural axons expressing DCC/FRA, but the axon cannot cross the midline to reach contralateral side, unless the attraction is suppressed (Serafini et al. 1996; Kidd et al. 1999). Subsequent in vivo studies, in *netrin-1*-deficient mice, exhibiting defects in commissure development, confirmed the requirement of Netrin-1 for spinal, corpus callosum, anterior, and hippocampal commissure development (Serafini et al. 1996). A role for Slit/SLT-1 in midline guidance came form studies of *slit-1* and *slit-2* mouse mutants, where callosal axons were unable to project across the midline and were deficient in all three major forebrain tracts, namely the corticofugal, callosal, and thalamocortical tracts (Bagri et al. 2002). In *Drosophila slit* mutant embryos, both ipsilateral and commissural axons enter the midline and never leave it (Rothberg et al. 1990; Battye et al. 1999; Kidd et al. 1999). In vertebrates, Slit achieves its function by signaling through Robo, and the Slit-2–Robo complex repels callosal axons away from the midline (Shu et al. 2003). Similarly, in *Drosophila*, Slit/SLT-1 signals via Roundabout (Robo) to prevent recrossing of axons supported by phenotypic analysis of *robo* mutants, where axons including ipsilateral freely cross and recross the midline multiple times (Seeger et al. 1993; Kidd et al. 1998). Later, two additional receptors for Slit were identified in *Drosophila*, Robo-2 and Robo-3, where Robo-2 and Robo-3 determine the position of longitudinal axons from the midline and Robo/Robo1 and Robo-2 prevented the commissural axons to linger in the midline (Rajgopalan et al. 2000). Based on these studies, it was proposed that Netrin/UNC-6 and FRA/DCC/UNC-40 attract the commissural axons to midline whereas Slit/SLT-1 and Robo/SAX-3 repel them to contralateral side and prevent recrossing.

In mammals, Rig-1/Robo-3 receptor, expressed in commissural axons, acts in midline guidance decisions probably by regulating sensitivity to Slit. Rig-1 functions by repressing the repulsive activity of any low level of Robo-1 present in precrossing axons. These results were validated by high level of Rig-1 expression before midline crossing and its downregulation after crossing. After crossing, Rig-1 is downregulated and Robo-1 and Robo-2 are upregulated in commissural GC, allowing their repulsion by the midline (Sabatier et al. 2004).

In *C. elegans*, the ventral midline is flanked on right and left sides by asymmetric VNC, containing bundle of longitudinal axon tracts (Fig. 2.3c), composed of two asymmetric bundles, a right bundle of 40 neurons and the left bundle of 4 neurons; in wild-type worms, the axons from the left bundle cross and extend on the contralateral side, whereas, the axons of right bundle remain ipsilateral. Phenotypic analysis of PVQ neuron (Fig. 2.3c) in *sax-3* mutants suggested that SAX-3/Robo functions to maintain the asymmetry of the VNC. In wild-type worms, the cell bodies of the two PVQ neurons, PVQR and PVQL, located in the tail of the worm and their axons travel anteriorly the length of the VNC on the right and left sides, respectively, whereas *sax-3* mutants exhibited repeated midline crossing by PVQ neuron observed in worms with *sra-6::GFP* transgene (Troemel et al. 1995), and the defects resembled the midline crossing defects in *Drosophila robo* mutants (Seeger et al. 1993; Zallen et al. 1998). Similar, crossing defects were also observed for HSN neuron (Fig. 2.3c). HSN neuron sends out a ventral process in response to UNC-6/Netrin acting through its receptor UNC-40/DCC/FRA to the VNC, followed by the extension of the process anteriorly without crossing the midline. In *sax-3* mutants, the longitudinal HSN process, cross and recross the ventral midline many times, suggesting a role for SAX-3 in HSN longitudinal process guidance (Zallen et al. 1998).

2.2.1 Wnt/Wnt and Ryk/Lin-18

Midline crossing in *Drosophila* deploys Wnt signaling in addition to the Netrin and Slit signaling pathways (Callahan et al. 1995; Yoshikawa et al. 2003). Wnts are secreted glycoproteins also called Wingless, the segment polarity gene of *Drosophila*, involved in the formation of body axis during embryonic development (Nusse and Varmus 1992). Wnts signal via Frizzled (Fz) receptors by binding to the CRDs (cysteine-rich domain) of Frizzled (Bhanot et al. 1996; Dann et al. 2001; Lin et al. 1997) and regulate key developmental processes including embryonic patterning, cell fate specification, and polarity determination (Ciani and Salinas 2005), mediating their responses through three signaling pathways including Wnt/β-catenin (canonical pathway), Wnt/planar cell polarity (non-canonical pathway), and Wnt/calcium pathway (Endo and Rubin 2007). Additionally, Wnts function during development of mammalian and fly CNS (Lyuksyutova et al. 2003; Liu et al. 2005). There are fourteen body segments in the *Drosophila*, each segment has an anterior and a posterior commissure, and axons cross the ventral midline once and choose one of the commissures (Callahan et al. 1995). The choice of commissure is determined by the Drl (derailed)/Ryk (receptor related to tyrosine kinase), an atypical receptor tyrosine kinase, expressed in the axons of the anterior commissure responding to Wnt5 protein, expressed mainly by axons in or close to the posterior commissure (Yoshikawa et al. 2003). Genetic analysis supported Wnt5 as a repulsive cue signaling via axons expressing Drl/Ryk, ensuring GC migration through the anterior commissure. In the absence of the Drl/Ryk receptor, axons were insensitive to Wnt5, resulting in improper commissure bundle formation, inappropriate crossing of the follower axons in the anterior commissure, and projecting inappropriately across the midline in between the commissures (Fradkin et al. 2004). Rescue experiments, expressing *WNT5* under a panneural promoter, confirmed the requirement of Wnt5 protein in commissure formation (Fradkin et al. 2004). Similarly, in mammals Wnt5a acts a repulsive ligand for Ryk receptor driving callosal axons towards the contralateral hemisphere after crossing the midline (Keeble et al. 2006).

Semaphorin protein, expressed at the midline and/or ventral tissues, have been shown to function as repulsive cue in midline guidance, genetic analysis of mice lacking neuropilin-2, a semaphorin receptor, exhibited midline guidance defects, confirming their role in this process (Zou et al. 1999). In worms, the VAB-1/Eph receptor also functions to prevent aberrant axon crossing at the ventral midline (Zallen et al. 1999).

2.3 Guidance Cues and Receptors Implicated in the Migration Along Anterior–Posterior Axis

Identification of molecules involved in axon guidance and cell migration along the AP or longitudinal axis has remained enigmatic, probably owing to the redundancy in DV and AP guidance and outgrowth signaling pathways. Section below discusses guidance molecules regulating GC migration along the AP axis of vertebrates and invertebrates.

2.3.1 Wnt and Fz

A number of studies have shown that Wnt proteins regulate AP guidance by regulating the polarity of migrating cells and neurons (Montcouquiol et al. 2006; Endo and Rubin 2007). Mammalian CNS is composed of sensory pathways that form ascending fibers relaying sensory stimuli to higher brain centers and descending fibers sending motor command and regulatory signals down from the brain. In the spinal cord, somatosensory fibers project anteriorly and motor pathways project posteriorly, making spinal cord an excellent model system for studying axon guidance along the AP axis. Mouse genome encodes for nineteen Wnt proteins and twelve Frizzled receptors (Miller 2002). Several of these genes are expressed in the mouse spinal cord gray matter, cupping the dorsal funiculus, in an anterior-to-posterior decreasing gradient along the cervical and thoracic cord. Biochemical studies found that descending neurons of the CST are repelled by Wnts and anti-Ryk antibodies blocked the posterior growth of CST axons, supporting the role of Wnts and Ryk receptor in posterior guidance of CST axons (Imondi and Thomas 2003; Liu et al. 2005). Similarly, commissural neurons in the mammalian dorsal spinal cord send axons ventrally toward the floor plate, where they cross the midline and turn anteriorly toward the brain. Several Wnt proteins stimulate the anterior extension of commissural axons after midline crossing. WNT4 mRNA, expressed in a decreasing anterior-to-posterior gradient in the floor plate, attracts postcrossing commissural axons. Commissural axons in mice lacking the Wnt receptor Fz3 (Frizzled3) displayed AP guidance defects after midline crossing implicating Wnt4 and Fz3 in anterior guidance of commissural axons (Lyuksyutova et al. 2003).

In *C. elegans*, Wnts are known to regulate cell fate, cell polarity, including axon outgrowth, and cell migration (Maloof et al. 1999; Whangbo and Kenyon 1999; Pan et al. 2006; Silkankova and Korswagen 2007; Zinovyeva et al. 2008). The worm genome encodes five Wnt proteins (EGL-20, LIN-44, CWN-1, CWN-2, and MOM-2) and four Frizzled receptors (MOM-5, CFZ-2, MIG-1, and LIN-17), along with the Ryk homologue LIN-18 (Pan et al. 2006). Wnts and their Fz receptors function in both cell and axon GC migrations in worms.

Q neuroblasts are born as bilaterally symmetric cells on the left (QL) and right (QR) sides in the posterior body region, each dividing asymmetrically and generating three neurons and two apoptotic cells. In the first larval stage (L1), QR and its descendents (AQR, SDQR, and AVM) migrate anteriorly and QL and its descendents (PQR, SDQL, and PVM) migrate posteriorly (Fig. 2.2a). Maloof et al. (1999) identified, EGL-20/Wnt, regulating the migrations of the Q neuroblast and its descendants by controlling the expression of Hox transcription factor MAB-5 (Male-ABnormal) in QL neuroblast through the canonical Wnt signaling pathway, migrating posteriorly into the tail towards the source of EGL-20/Wnt. Studies conducted by Whangbo and Kenyon (1999), implicated LIN-17/Fz as the receptor since due to higher concentrations of EGL-20/Wnt, the QL neuron expressed the Wnt pathway proteins such as LIN-17/Fz, BAR-1/β-catenin, and MIG-5/Dishevelled, resulting in posterior migration while relatively lower EGL-20/Wnt concentrations and absence of the Wnt pathway proteins resulted in QR migration in the anterior direction. Later, genetic analysis showed that EGL-20 functions in QL migration by signaling through MIG-1/Fz and LIN-17/Fz proteins (Forrester et al. 2004; Harris et al. 1996; Maloof et al. 1999).

Wnts were also implicated in several other cell migrations along the AP axis of the worm including CAN and ALM neurons (Fig. 2.2c), which undergo posterior migration near the middle of the worm body and HSNs and BDUs (Fig. 2.2c), migrating anteriorly during development. Loss-of-function mutations in Wnts, including *egl-20*, *cwn-1*, and *cwn-2*, resulted in migration defects in several neurons including ALM, BDU, CAN, HSN, and QR evidenced by phenotypic analysis of *cwn-1 cwn-2* double mutants; moreover, studies supported CWN-2 signaling via CFZ-2/Fz (Zinovyeva and Forrester 2005).

EGL-20, CWN-1, and LIN-44, expressed by a group of epidermal and muscle cells located in the tail of the worm together with a fourth Wnt, MOM-2, function redundantly to repel HSN neurons and promote their anterior migrations from the tail region to midbody (Fig. 2.2c), where EGL-20 carried out its repulsive function via Frizzled, MIG-1 and MOM-5 (Pan et al. 2006). Besides, Wnt signaling has been associated with GC guidance of the anterior processes of AVM and PVM neurons (Fig. 2.1a), phenotypic analysis of various single and double mutants supported the requirement of CWN-1 and EGL-20 Wnts and MIG-1 and MOM-5 Frizzled in the GC migrations (Pan et al. 2006).

Wnts and Frizzled have been also associated with polarity establishment and process outgrowth in ALM and PLM neurons (Fig. 2.1a) along the AP axis of the worm. PLM, which normally sends an anterior process in wild-type worm, extended a posterior process in *lin-44* and *lin-17* loss-of-function mutants. Two Wnts, including CWN-1 and EGL-20 were associated with ALM polarity reversal (Hilliard and Bargmann 2006).

2.3.2   EGL-17/FGF and EGL-15/FGFR

The fibroblast growth factor (FGF) and its receptor FGFR, receptor tyrosine kinase pathway, have been implicated in numerous cell migrations during development, in both vertebrates and invertebrates (Chen and Stern 1998; Metzger and Krasnow 1999; Hogan 1999). In the *Drosophila* trachea and the vertebrate lung, FGF directs branching and morphogenesis, a process that includes both cell migrations and cell shape changes (Chu et al. 2013; Bae et al. 2012). In *Drosophila, breathless (btl)*, homologue of FGF, signals via *branchless (bln)*, homologue of FGFR, for normal migration of tracheal cells (Glazer and Shilo 1991; Samakovlis et al. 1996). In vertebrates, FGF and FGFR signaling has been implicated in neuronal migration, neurite outgrowth, morphogenesis, and cell fate specification (Hu et al. 2013; Neiiendam et al. 2004; Hossain and Morest 2000). In *C. elegans*, FGF–FGFR signaling pathway has been implicated in migrations of SMs and canal-associated neurons (CANs) along the AP axis. SM cells are born in the posterior of the animal during the first larval stage, and in hermaphrodite, they migrate anteriorly to the central gonad and developing vulva, generating uterine and vulval musculature (Fig. 2.2d). Proper positioning of SM requires an FGF-like ligand encoded by the *egl-17* gene, expressed in somatic gonadal cells, acting as attractive cue, and a receptor belonging to the FGF receptor subfamily, encoded by *egl-15*, expressed in SMs (Burdine et al. 1997, 1998; DeVore et al. 1995), constituting the GDA mechanism of SM migration. Mutations in either *egl-17* or *egl-15* result in SM migration defects. Recently, *egl-17* and *egl-15* signaling pathway has been implicated in CAN neuron (Fig. 2.2c) migration, where EGL-17 acts as a repulsive guidance cue evidenced by repelling CAN neurons from sources of ectopic expression of EGL-17, possibly mediating its effect through adaptor protein, SEM-5/GRB-2 (Fleming et al. 2005).

2.3.3   SLT-1/Slit and SAX-3/Robo

SLT-1 functions in multiple guidance pathways including midline, DV, and AP guidance decision. Along the longitudinal axis, it functions in guidance of CAN and ALM neurons (Fig. 2.2c). In embryos, SLT-1 is expressed at high levels in the anterior epidermis, providing guidance for posterior CAN migration to midbody region. CAN migration defects were observed in SLT-1 loss-of-function worms with *ceh-23::GFP* transgene (Zallen et al. 1999; Hao et al. 2001; Yu et al. 2002). SLT-1 functions through its receptor SAX-3/Robo in CAN migration evidenced by CAN migration defects in *slt-1* mutant worms with *myo-3::SLT-1* transgene, where SLT-1 is expressed in muscles. Mutations in *slt-1* and *sax-3* also affected the posterior migrations of ALM neurons (Fig. 2.2c) supported by phenotypic analysis of mutant worms. Longitudinal migration of GCs in VNC (Fig. 2.3c), are also affected by mutation in *sax-3*, phenotypic analysis of VNC structure in *sax-3* mutants visualized with *glr-1::GFP* transgene (Maricq et al. 1995), revealed premature axon termination (Zallen et al. 1998), supporting SAX-3 requirement in VNC axon elongation.

2.3.4 MIG-13

mig-13 gene encodes a transmembrane protein, composed of LDL (low-density lipoprotein) receptor repeats and CUB (for complement C1r/C1s, Uegf, Bmp1) domain, affecting the anterior migration of QR and its descendents (QR.x) (Fig. 2.2a), acting cell autonomously. MIG-13 is expressed in the anterior and central regions of the worm under the act of Hox gene. Loss of MIG-13 activity shifts the QR.x (QR and its descendent) posteriorly, whereas increasing the level of MIG-13 shifts them anteriorly in a dose-dependent manner (Sym et al. 1999; Wang et al. 2013). Moreover, recently, it has been shown that LIN-39 establishes QR anterior polarity by binding to the *mig-13* promoter and enhancing its expression, whereas expression of *lin-39* is inhibited in QL.x (QL and its descendent), thereby providing a link between *mig-13* expression and anterior migration of QR.x. MAB-5, a Wnt pathway protein, implicated in posterior migration of QL, inhibits QL anterior polarity by associating with the *lin-39* promoter and down-regulating *lin-39* and thereby *mig-13* expression (Wang et al. 2013).

2.3.5 UNC-53/ NAVs

unc-53 gene, which regulates the AP directional migrations (Hedgecock et al. 1987), encodes an actin-binding scaffolding protein with calponin homology actin-binding domain, AAA domain, coiled-coil regions, and two polyproline-rich region (Stringham et al. 2002). Hypomorphic alleles of *unc-53* display guidance defects in the anterior migration of the SM (Chen et al. 1997) and the process outgrowth of mechanosensory neurons (Fig. 2.1a), ALN and PLN (Fig. 2.2b) neurons, motor neurons (Fig. 2.1d), and the excretory canals (Fig. 2.3a) (Hekimi and Kershaw 1993; Stringham et al. 2002). Three mammalian UNC-53 homologues, NAV-1/UNC53H1, NAV-2/UNC53H2 and NAV-3/UNC53H3 (neuron navigator-1, 2, 3) coined navigators for their role in axon guidance, have been identified (Merrill et al. 2002; Maes et al. 2002), shown to be involved in neurite induction, supporting the functional conservation of UNC-53/NAV in both vertebrates and invertebrates.

2.3.6 VAB-8

VAB-8 (*variable ab*normal morphology), a kinesin-like protein, functioning primarily in posterior GC migration, was identified in a screen for mutants affecting CAN neuron migration based on the "withered tail" phenotype, which is associated with abnormal CAN migration (Manser and Wood 1990). VAB-8 affects GC migration both during embryogenesis and during larval stages, and mutations in *vab-8* disrupt

Table 2.1   Guidance molecules involved in growth cone migration and axon outgrowth

Cue/Receptor complex	Organism	References
Dorsal–ventral (DV) guidance molecules		
UNC-6/UNC-40	Nematode	Chan et al. (1996), Wadsworth et al. (1996)
Netrin/Neogenin/DCC	Vertebrates	Serafini et al. (1994), Keino-Masu et al. (1996)
SLT-1/SAX-3	Nematode	Brose et al. (1999), Hao et al. (2001)
SLT-1/EVA-1	Nematode	Fujisawa et al. (2007)
UNC-6/UNC-5	Nematode	Leung-Hagesteijn et al. (1992), Wadsworth et al. (1996)
UNC-6/UNC-5/Rcm	Vertebrate	Ackerman et al. (1997), Ackerman and Knowles (1998), Finger et al. (2002)
UNC-129/UNC-5	Nematode	Colavita et al. (1998)
TGF-β/TGFR-β	Vertebrates	Augsburger et al. (1999), Dillon et al. (2007)
UNC-6/ENU-3	Nematode	Yee et al. (2011)
Midline guidance molecules		
SLIT-1/Robo	Vertebrates	Rothberg et al. (1990), Seeger et al. (1993), Kidd et al. (1998)
SLT-1/SAX-3	Nematodes	Zallen et al. (1998)
Wnt5/Drl/Ryk	Fly	Callahan et al. (1995), Yoshikawa et al. (2003), Fradkin et al. (2004)
Anterior–posterior (AP)/Longitudinal guidance molecules		
Wnt/Ryk	Vertebrates	Imondi and Thomas (2003), Liu et al. (2005), and Dickson and Gilestro (2006)
Wnt/Fz3	Vertebrates	Lyuksyutova et al. (2003)
Wnts/Fz	Nematode	Maloof et al. (1999), Zinovyeva and Forrester (2005), Pan et al. (2006), Hilliard and Bargmann (2006)
FGF/FGFR	Vertebrates	Hu et al. (2013), Neiiendam et al. (2004), Hossain and Morest (2000)
EGL-17/EGL-15	Nematodes	Burdine et al. (1997, 1998), Fleming et al. (2005)
SLT-1/SAX-3	Nematodes	Zallen et al. (1998), Zallen et al. (1999), Hao et al. (2001), Yu et al. (2002)
UNC-53	Nematodes	Hekimi and Kershaw (1993), Stringham et al. (2002)
VAB-8	Nematode	Wightman et al. (1996)
MIG-13	Nematodes	Sym et al. (1999), Wang et al. (2013)

fourteen of the seventeen posteriorly directed migrations (Wightman et al. 1996), a role further validated by expression of VAB-8 in the mechanosensory neuron (Fig. 2.1a) under the control of *mec-4* promoter, causing rerouting of ALM anterior process toward the tail of the worm. Besides, *vab-8* regulates the posterior migration of ALM during embryogenesis (Wolf et al. 1998). Genetic analysis implicated VAB-8 and UNC-73B function in EC process outgrowth (Fig. 2.2b), working in a parallel pathway to UNC-53 and UNC-73E (Marcus-Gueret et al. 2012).

Table 2.1 provides a list of Guidance molecules involved in GC steering during embryogenesis discussed in the above sections, and a schematic is depicted in Fig. 2.4.

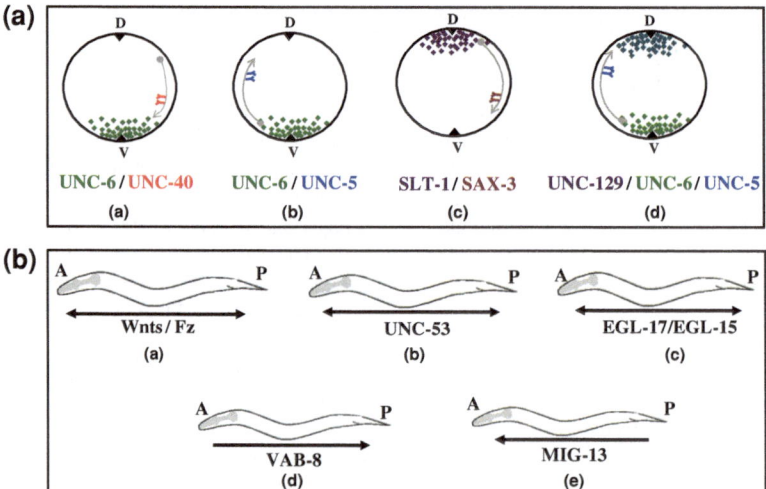

Fig. 2.4 Graphic of guidance molecules in dorsal–ventral and anterior–posterior growth cone migrations and guidance in *C. elegans*: **a** Guidance molecules involved in dorsal–ventral (*DV*) migration in worms. Cross section of the worm depicting dorsal (*D*) and ventral (*V*) sides. *a* UNC-6/UNC-40, where UNC-6 is concentrated ventrally and attracts GCs expressing UNC-40 receptor. *b* UNC-6/UNC-5, ventral source of UNC-6 repels GCs expressing UNC-5 dorsally. *c* SLT-1/SAX-3, SLT-1 is concentrated dorsally and repels growth cones expressing SAX-3 receptor ventrally. *d* UNC-129/UNC-5, attract growth cones expressing UNC-5 receptor repelled by ventrally located UNC-6. **b** Lateral view of the worm depicting anterior (*A*) and posterior (*P*) growth cone migrations. Guidance molecules involved in AP guidance include the following: *a* Wnts/Fz, Wnts acting through the Frizzled (*Fz*) receptors guiding growth cones along the AP axis. *b* EGL-15/EGL-17, EGL-15 a secreted cue binds EGL-17 receptor, regulating GC migrations along the longitudinal axis. *c* UNC-53, a cytoplasmic scaffolding protein steering GCs in both anterior and posterior directions. *d* MIG-13, a transmembrane protein regulating anterior GC migrations. *e* VAB-8, a kinesin-like protein, regulating posterior GC migrations

Some of the molecules involved in axon guidance that is outside the scope of discussion of this book include semaphorins, a family of secreted proteins signaling through plexins (Kruger et al. 2005) and ephrins acting through their Eph receptors (Flanagan 2006; Quinn and Wadsworth 2006; Mohamed and Chin-Sang 2006).

References

Ackerman SL, Knowles BB (1998) Cloning and mapping of the UNC5C gene to human chromosome 4q21-q23. Genomics 52(2):205-208

Ackerman SL, Kozak LP, Przyborski SA, Rund LA, Boyer BB, Knowles BB (1997) The mouse rostral cerebellar malformation gene encodes an UNC-5-like protein. Nature 386(6627):838–842

Adler CE, Fetter RD, Bargmann CI (2006) UNC-6/Netrin induces neuronal asymmetry and defines the site of axon formation. Nat. Neurosci. 9:511–518

Alexander F (2000) Neuroblastoma. Urol Clin North Am 3:383–392 vii

Alexander M, Selman G, Seetharaman A, Chan KK, D'Souza SA, Byrne AB, Roy PJ (2010) MADD-2, a homolog of the Opitz syndrome protein MID1, regulates guidance to the midline through UNC-40 in *Caenorhabditis elegans*. Dev. Cell. 18(6):961–972

Augsburger A, Schuchardt A, Hoskins S, Dodd J, Butler S (1999) BMPs as mediators of roof plate repulsion of commissural neurons. Neuron 24:127–141

Bae YK, Trisnadi N, Kadam S, Stathopoulos A (2012) The role of FGF signaling in guiding coordinate movement of cell groups: guidance cue and cell adhesion regulator? Cell Adh Migr 6(5):397–403

Bagri A, Marín O, Plump AS, Mak J, Pleasure SJ, Rubenstein JL, Tessier-Lavigne M (2002) Slit proteins prevent midline crossing and determine the dorsoventral position of major axonal pathways in the mammalian forebrain. Neuron 33:233–248

Barallobre MJ, Del Rio JA, Alcantara S, Borrell V, Aguado F, Ruiz M, Carmona MA, Martin M, Fabre M, Yuste R, Tessier-Lavigne M, Soriano E (2000) Aberrant development of hippocampal circuits and altered neural activity in Netrin 1-deficient mice. Development 127:4797–4810

Battye R, Stevens A, Jacobs JR (1999) Axon repulsion from the midline of the *Drosophila* CNS requires Slit function. Development 126:2475–2481

Bhanot P, Brink M, Samos CH, Hsieh JC, Wang Y, Macke JP, Andrew D, Nathans J, Nusse R (1996) A new member of the frizzled family from *Drosophila* functions as a Wingless receptor. Nature 382:225–230

Braisted JE, Catalano SM, Stimac R, Kennedy TE, Tessier-Lavigne M, Shatz CJ, O'Leary DD (2000) Netrin-1 promotes thalamic axon growth and is required for proper development of the thalamocortical projection. J. Neurosci. 20:5792–5801

Branda CS, Stern MJ (2000) Mechanisms controlling sex myoblast migration in *Caenorhabditis elegans* hermaphrodites. Dev Biol 226(1):137–151

Brose K, Bland KS, Wang KH, Arnott D, Henzel W, Goodman CS, Tessier-Lavigne M, Kidd T (1999) Slit proteins bind Robo receptors and have an evolutionarily conserved role in repulsive axon guidance. Cell 96(6):795–806

Burdine RD, Branda CS, Stern MJ (1998) EGL-17(FGF) expression coordinates the attraction of the migrating sex myoblasts with vulval induction in *C. elegans*. Development 125:1083–1093

Burdine RD, Chen EB, Kwok SF, Stern MJ (1997) *egl-17* encodes an invertebrate fibroblast growth factor family member required specifically for sex myoblast migration in *Caenorhabditis elegans*. Proc Natl Acad Sci USA 94:2433–2437

Callahan CA, Muralidhar MG, Lundgren SE, Scully AL, Thomas JB (1995) Control of neuronal pathway selection by a *Drosophila*-tyrosine kinase family member. Nature 376:171–174

Ciani L, Salinas PC (2005) Wnts in the vertebrate nervous system: from patterning to neuronal connectivity. Nat. Rev. Neurosci. 6:351–362

Chalfie M, Sulston JE, White JG, Southgate E, Thomson JN, Brenner S (1985) The neural circuit for touch sensitivity in *Caenorhabditis elegans*. J Neurosci. 5(4):956–964

Chalfie M, Sulston J (1981) Developmental genetics of the mechanosensory neurons of *Caenorhabditis elegans*. Dev Biol. 82(2):358–370

Chalfie M, Tu Y, Euskirchen G, Ward WW, Prasher DC (1994) Green fluorescent protein as a marker for gene expression. Science 263:802–805

Chan SS, Zheng H, Su MW, Wild R, Killeen MT, Hedgecock EM, Culotti JG (1996) UNC-40, a *C. elegans* homolog of DCC (deleted in colorectal cancer), is required in motile cells responding to UNC-6 Netrin cues. Cell 87:187–195

Chen EB, Branda CS, Stern MJ (1997) Genetic enhancers of *sem-5* define components of the gonad-independent guidance mechanism controlling sex myoblast migration in *Caenorhabditis elegans* hermaphrodites. Dev Biol 182(1):88–100

Chen EB, Stern MJ (1998) Understanding cell migration guidance: lessons from sex myoblast migration in *C. elegans*. Trends Genetics 14:322–327

Chu WC, Lee YM, Henry Sun Y (2013) FGF/FGFR signal induces trachea extension in the drosophila visual system. PLoS ONE 8(8):e73878

Colamarino SA, Tessier-Lavigne M (1995) The axonal chemoattractant Netrin-1 is also a chemorepellent for trochlear motor axons. Cell 81:621–629

Colavita A, Culotti JG (1998) Suppressors of ectopic UNC-5 growth cone steering identify eight genes involved in axon guidance in *Caenorhabditis elegans*. Dev Biol 194:72–85

Colavita A, Krishna S, Zheng H, Padgett RW, Culotti JG (1998) Pioneer axon guidance by UNC-129, a *C. elegans* TGF-beta. Science 28:706–709

Dann CE, Hsieh JC, Rattner A, Sharma D, Nathans J, Leahy DJ (2001) Insights into Wnt binding and signaling from the structures of two frizzled cysteine-rich domains. Nature 412:86–90

DeVore DL, Horvitz HR, Stern MJ (1995) An FGF receptor signaling pathway is required for the normal cell migrations of the sex myoblasts in *C. elegans* hermaphrodites. Cell 83:611–620

Dickson BJ, Gilestro GF (2006) Regulation of commissural axon pathfinding by Slit and its Robo receptors. Annu. Rev. Cell Dev. Biol. 22:651–675

Dillon AK, Jevince AR, Hinck L, Ackerman SL, Lu X, Tessier-Lavigne M, Kaprielian Z (2007) UNC5C is required for spinal accessory motor neuron development. Mol Cell Neurosci 35:482–489

Endo Y, Rubin JS (2007) Wnt signaling and neurite outgrowth: insights and questions. Cancer Science 98:1311–1317

Fearon ER, Cho KR, Nigro JM, Kern SE, Simons JW, Rup- pert JM, Mamilton SR, Preisinger AC, Thomas G, Kinzler KW, Vogelstein B (1990) Identification of a chromosome 18q gene that is altered in colorectal cancers. Science 247:49–56

Finger JH, Bronson RT, Harris B, Johnson K, Przyborski SA, Ackerman SL (2002) The Netrin 1 Receptors *Unc5h3* and *Dcc* are necessary at multiple choice points for the guidance of corticospinal tract axons. J Neurosci 22(23):10346–10356

Flanagan JG (2006) Neural map specification by gradients. Curr Opin Neurobiol 16:59–66

Fleming TC, Wolf FW, Garriga G (2005) Sensitized genetic backgrounds reveal a role for *C. elegans* FGF EGL-17 as a repellent for migrating CAN neurons. Development 132:4857–4867

Forrester WC, Perens E, Zallen JA, Garriga G (1998) Identification of *Caenorhabditis elegans* genes required for neuronal differentiation and migration. Genetics 148(1):151--165

Fradkin LG, van Schie M, Wouda RR, de Jong A, Kamphorst JT, Radjkoemar-Bansraj M, Noordermeer JN (2004) The *Drosophila* Wnt5 protein mediates selective axon fasciculation in the embryonic central nervous system. Dev Biol 272:362–375

Fujisawa K, Wrana JL, Culotti JG (2007) The slit receptor EVA-1 coactivates a SAX-3/robo mediated guidance signal in *C. elegans*. Science 317:1934–1938, Erratum in: Science 318:570

Guijarro P, Simó S, Pascual M, Abasolo I, Del Río JA, Soriano E (2006) Netrin1 exerts a chemorepulsive effect on migrating cerebellar interneurons in a Dcc-independent way. Mol Cell Neurosci 33(4):389–400

Glazer L, Shilo BZ (1991) The *Drosophila* FGF-R homolog is expressed in the embryonic tracheal system and appears to be required for directed tracheal cell extension. Genes Dev 5:697–705

Hamelin M, Zhou Y, Su MW, Scott IM, Culotti JG (1993) Expression of the UNC-5 guidance receptor in the touch neurons of *C. elegans* steers their axons dorsally. Nature 364(6435):327–330

Hao JC, Yu TW, Fujisawa K, Culotti JG, Gengyo-Ando K, Mitani S, Moulder G, Barstead R, Tessier-Lavigne M, Bargmann CI (2001) *C. elegans* Slit acts in midline, dorsal–ventral, and anterior–posterior guidance via the SAX-3/Robo receptor. Neuron 32:25–38

Harris J, Honigberg L, Robinson N, Kenyon C (1996) Neuronal cell migration in *C. elegans*: regulation of Hox gene expression and cell position. Development 122:3117–3131

Hedgecock EM, Culotti JG, Hall DH (1990) The *unc-5*, *unc-6*, and *unc-40* genes guide circumferential migrations of pioneer axons and mesodermal cells on the epidermis in *C. elegans*. Neuron 4:61–85

Hedgecock EM, Culotti JG, Hall DH, Stern BD (1987) Genetics of cell and axon migrations in *Caenorhabditis elegans*. Development 100:365–382

Hekimi S, Kershaw D (1993) Axonal Guidance defects in a *Caenorhabditis elegans* mutant reveal cell-extrinsic determinants of neuronal morphology. The Journal of Neuroscience 13(10):4254–4271

Hilliard MA, Bargmann CI (2006) Wnt signals and frizzled activity orient anterior–posterior axon outgrowth in *C. elegans*. Dev. Cell 10:379–390

Hogan BLM (1999) Morphogenesis. Cell 96:225–233

Hong K, Hinck L, Nishiyama M, Poo MM, Tessier-Lavigne M, Stein E (1999) A ligand-gated association between cytoplasmic domains of UNC5 and DCC family receptors converts Netrin-induced growth cone attraction to repulsion. Cell 97:927–941

Hossain WA, Morest DK (2000) Fibroblast growth factors (FGF-1, FGF-2) promote migration and neurite growth of mouse cochlear ganglion cells in vitro: immunohistochemistry and antibody perturbation. J Neurosci Res 62(1):40–55

Hu Y, Poopalasundaram S, Graham A, Bouloux PM (2013) GnRH neuronal migration and olfactory bulb neurite outgrowth are dependent on FGF receptor 1 signaling, specifically via the PI3K p110α isoform in chick embryo. Endocrinology 154(1):388–399

Imondi R, Thomas JB (2003) Neuroscience. The ups and downs of Wnt signaling. Science 302:1903–1904

Itoh A, Miyabayashi T, Ohno M, Sakano S (1998) Cloning and expressions of three mammalian homologues of Drosophila slit suggest possible roles for slit in the formation and maintenance of the nervous system. Brain Res. Mol. Brain Res. 62:175–186

Keeble TR, Halford MM, Seaman C, Kee N, Macheda M, Anderson RB, Stacker SA, Cooper HM (2006) The Wnt receptor Ryk is required for Wnt5a-mediated axon guidance on the contralateral side of the corpus callosum. J Neuroscience 26:5840–5848

Keino-Masu K, Masu M, Hinck L, Leonardo ED, Chan SS, Culotti JG, Tessier- Lavigne M (1996) Deleted in colorectal cancer (DCC) encodes a Netrin receptor. Cell 87:175–185

Keleman K, Dickson BJ (2001) Short- and long- range repulsion by the *Drosophila* Unc5 Netrin receptor. Neuron 32:605–617

Keleman K, Rajagopalan S, Cleppien D, Teis D, Paiha K, Huber LA, Technau, GM, Dickson BJ (2002) Comm sorts robo to control axon guidance at the Drosophila midline. Cell 110:415–427

Keleman K, Ribeiro C, Dikson BJ (2005) Comm function in commissural axon guidance: cell-autonomous sorting of Robo in vivo. Nat. Neurosci. 8:156–163

Kennedy TE, Serafini T, de la Torre JR, Tessier-Lavigne M (1994) Netrins are diffusible chemotropic factors for commissural axons in the embryonic spinal cord. Cell 78:425–435

Kidd T, Bland KS, Goodman CS (1999) Slit is the midline repellent for the robo receptor in Drosophila. Cell 96(6):785–794

Kidd T, Brose K, Mitchell KJ, Fetter RD, Tessier-Lavigne M, Goodman CS, Tear G (1998a) Roundabout controls axon crossing of the CNS midline and defines a novel subfamily of evolutionarily conserved guidance receptors. Cell 92(2):205–215

Kidd T, Russell C, Goodman CS, Tear G (1998b) Dosage-sensitive and complementary functions of roundabout and commissureless control axon crossing of the CNS midline. Neuron 20:25–33

Kim S, Ren X-C, Fox E, Wadsworth WG (1999) SDQR migrations in *Caenorhabditis elegans* are controlled by multiple guidance cues and changing responses to netrin UNC-6. Development 126:3881–3890

Kolodziej PA, Timpe LC, Mitchell KJ, Fried SR, Goodman CS, Jan LY, Jan YN (1996) Frazzled encodes a Drosophila member of the DCC immunoglobulin subfamily and is required for CNS and motor axon guidance. Cell 87:197–204

Kruger RP, Aurandt J, Guan KL (2005) Semaphorins command cells to move. Nat Rev Mol Cell Biol 6:789–800

Leonardo ED, Hinck L, Masu M, Keino-Masu K, Ackerman SL, Tessier-Lavigne M (1997) Vertebrate homologues of *C. elegans* UNC-5 are candidate netrin receptors. Nature 386(6627):833–838

Leung-Hagesteijn C, Spence AM, Stern BD, Zhou Y, Su MW, Hedgecock EM, Culotti JG (1992) UNC-5, a transmembrane protein with immunoglobulin and thrombospondin type 1 domains, guides cell and pioneer axon migrations in *C. elegans*. Cell 71(2):289–299

Lin L, Rao Y, Isacson O (2005) Netrin-1 and slit-2 regulate and direct neurite growth of ventral midbrain dopaminergic neurons. Mol. Cell Neurosci. 28:547–555

Lin K, Wang S, Julius MA, Kitajewski J, Moos M, Luyten FP (1997) The cysteine-rich frizzled domain of Frzb-1 is required and sufficient for modulation of Wnt signaling. Proc. Natl. Acad. Sci. U. S. A. 94:11196–11200

Liu Y, Shi J, Lu CC, Wang ZB, Lyuksyutova AI, Song XJ, Zou Y (2005) Ryk-mediated Wnt repulsion regulates posterior-directed growth of corticospinal tract. Nat. Neurosci. 8:1151–1159

Lyuksyutova AI, Lu CC, Milanesio N, King LA, Guo N, Wang Y, Nathans J, Tessier-Lavigne M, Zou Y (2003) Anterior–posterior guidance of commissural axons by Wnt-frizzled signaling. Science 302:1984–1988

Maduro M, Pilgrim D (1995) Identification and cloning of *unc-119*, a gene expressed in the *Caenorhabditis elegans* nervous system. Genetics 141:977–988

Maes T, Barcelo A, Buesa C (2002) Neuron navigator: a human gene family with homology to *unc-53*, a cell guidance gene from *Caenorhabdtis elegans*. Genomics 80:21–30

Maloof JN, Whangbo J, Harris JM, Jongeward GD, Kenyon C (1999) A Wnt signaling pathway controls *hox* gene expression and neuroblast migration in *C. elegans*. Development 126:37–49

Manser J, Wood WB (1990) Mutations affecting embryonic cell migrations in *Caenorhabditis elegans*. Dev. Genet. 11(1):49–64

Marcus-Gueret N, Schmidt KL, Stringham EG (2012) Distinct cell guidance pathways controlled by the Rac and Rho GEF domains of UNC-73/TRIO in *Caenorhabditis elegans*. Genetics 190:129–142

Maricq AV, Peckol E, Driscoll M, Bargmann CI (1995) Mechanosensory signaling in *C. elegans* mediated by the GLR-1 glutamate receptor. Nature 378:78–81

Merz DC, Zheng H, Killeen MT, Krizus A, Culotti JG (2001) Multiple signaling mechanisms of the UNC-6/Netrin receptors UNC-5 and UNC-40/DCC in vivo. Genetics 158(3):1071–1080

Métin C, Deléglise D, Serafini T, Kennedy TE, Tessier-Lavigne M (1997) A role for netrin-1 in the guidance of cortical efferents. Development 124(24):5063–5074

Metzger RJ, Krasnow MA (1999) Genetic control of branching morphogenesis. Science 284:1635–1639

Merrill RA, Plum LA, Kaiser ME, Clagett-Dame M (2002) A mammalian homolog of *unc-53* is regulated by all-trans retinoic acid in neuroblastoma cells and embryos. Proc Natl Acad Sci U.S.A 99(6):3422–3427

Miller JR (2002) The Wnts. Genome Biol 3:3001

Mitchell KJ, Doyle JL, Serafini T, Kennedy TE, Tessier-Lavigne M, Goodman CS, Dickson BJ (1996) Genetic analysis of Netrin genes in Drosophila: Netrins guide CNS commissural axons and peripheral motor axons. Neuron 17:203–215

Mohamed AM, Chin-Sang ID (2006) Characterization of loss-of-function and gain-of-function Eph receptor tyrosine kinase signaling in *C. elegans* axon targeting and cell migration. Dev Biol 290:164–176

Montcouquiol M, Crenshaw EB III, Kelley MW (2006) Noncanonical Wnt signaling and neural polarity. Annu Rev Neurosci 29:363–386

Neiiendam JL, Køhler LB, Christensen C, Li S, Pedersen MV, Ditlevsen DK, Kornum MK, Kiselyov VV, Berezin V, Bock E (2004) An NCAM-derived FGF-receptor agonist, the FGL-peptide, induces neurite outgrowth and neuronal survival in primary rat neurons. J Neurochem 91(4):920–935

Nusse R, Varmus HE (1992) Wnt genes. Cell 26(7):1073–1087

Nüsslein-Volhard C, Wieschaus E, Kluding H (1984) Mutations affecting the pattern of the larval cuticle in *Drosophila melanogaster* I. Zygotic loci on the second chromosome Roux's Arch. Dev Biol 193:267–282

Pan CL, Howell JE, Clark SG, Hilliard M, Cordes S, Bargmann CI, Garriga G (2006) Multiple Wnts and frizzled receptors regulate anteriorly directed cell and growth cone migrations in *Caenorhabditis elegans*. Dev. Cell 10:367–377

Qin H, Powell-Coffman JA (2004) The *Caenorhabditis elegans* aryl hydrocarbon receptor, AHR-1, regulates neuronal development. Dev Biol 270:64–75

Quinn CC, Wadsworth WG (2006) Axon guidance: ephrins at WRK on the midline. Curr. Biol. 16:R954–R955

Quinn CC, Pfeil DS, Chen E, Stovall EL, Harden MV, Gavin MK, Forrester WC, Ryder EF, Soto MC, Wadsworth WG (2006) UNC-6/Netrin and SLT-1/slit guidance cues orient axon outgrowth mediated by MIG-10/RIAM/lamellipodin. Curr Biol 16:845–853

Rajagopalan S, Vivancos V, Nicolas E, Dickson BJ (2000) Selecting a longitudinal pathway: Robo receptors specify the lateral position of axons in the Drosophila CNS. Cell 103:1033–1045

Rothberg JM, Hartley DA, Walther Z, Artavanis-Tsakonas S (1988) *slit:* an EGF-homologous locus of *D. melanogaster* involved in the development of the embryonic central nervous system. Cell 55:1047–1059

Rothberg JM, Jacobs JR, Goodman CS, Artavanis-Tsakonas S (1990) *Slit*: an extracellular protein necessary for development of midline glia and commissural axon pathways contains both EGF and LRR domains. Genes Dev 4:2169–2187

Sabatier C, Plump AS, Ma L, Brose K, Tamada A, Murakami F, Lee EV, Tessier-Lavigne M (2004) The divergent robo family protein *rig-1*/robo3 is a negative regulator of Slit responsiveness required for midline crossing by commissural axons. Cell 117:157–169

Samakovlis C, Hacohen N, Manning G, Sutherland D, Guillemin K, Krasnow MA (1996) Branching morphogenesis of the *Drosophila* tracheal system occurs by a series of morphologically distinct but genetically coupled branching events. Development 122:1395–1407

Seeger M, Tear G, Ferres-Marco D, Goodman CS (1993) Mutations affecting growth cone guidance in Drosophila: genes necessary for guidance toward or away from the midline. Neuron 10:409–426

Seetharaman A, Selman G, Puckrin R, Barbier L, Wong E, D'Souza SA, Roy PJ (2011) MADD-4 is a secreted cue required for midline-oriented guidance in *Caenorhabditis elegans*. Dev Cell. 21(4):669–680

Serafini T, Kennedy TE, Galko MJ, Mirzayan C, Jessell TM, Tessier-Lavigne M (1994) The Netrins define a family of axon outgrowth-promoting proteins homologous to *C. elegans* UNC-6. Cell 78:409–424

Serafini T, Colamarino SA, Leonardo ED, Wang H, Beddington R, Skarnes WC, Tessier-Lavigne M (1996) Netrin-1 is required for commissural axon guidance in the developingvertebrate nervous system. Cell 87:1001–1014

Shewan D, Dwivedy A, Anderson R, Holt CE (2002) Age-related changes underlie switch in Netrin-1 responsiveness as growth cones advance along visual pathway. Nat Neurosci 5:955–962

Shu T, Sundaresan V, McCarthy M, Richards LJ (2003) Slit2 guides both precrossing and postcrossing callosal axons at the midline in vivo. J Neurosci 23:8176–8184

Silhankova M, Korswagen HC (2007) Migration of neuronal cells along the anterior–posterior body axis of *C. elegans*: Wnts are in control. Curr. Opin. Genet. Dev. 17:320–325

Stern MJ, Horvitz HR (1991) A normally attractive cell interaction is repulsive in two *C. elegans* mesodermal cell migration mutants. Development 113:797–803

Stein E, Tessier-Lavigne M (2001) Hierarchical organization of guidance receptors: silencing of Netrin attraction by Slit through a Robo/DCC receptor complex Science 291:1928–1938

Stringham E, Pujol N, Vanderchove J, Bogaert T (2002) unc-53 controls longitudinal migration in *C. elegans* Development 129:3367–3379

Sulston JE, Horvitz HR (1977) Post-embryonic cell lineages of the nematode *Caenorhabditis elegans*. Dev. Biol. 56:110–156

Sulston JE, Schierenberg E, White JG, Thomson JN (1983) The embryonic cell lineage of the nematode *Caenorhabditis elegans*. Dev Biol 100:64–119

Sym M, Robinson N, Kenyon C (1999) MIG--13 positions migrating cells along the anteroposterior body axis of *C. elegans*. Cell 98(1):25–36

Tessier-Lavigne M, Goodman CS (1996) The molecular biology of axon guidance. Science 274: 1123–1133

Thomas JH, Stern MJ, Horvitz HR (1990). Cell interactions coordinate the development of the *C. elegans* egg-laying. Cell 62:1041–1052

Troemel ER, Chou JH, Dwyer ND, Colbert HA, Bargmann CI (1995) Divergent seven transmembrane receptors are candidate chemosensory receptors in *C. elegans*. Cell 83:207–218

Wadsworth WG, Bhatt H, Hedgecock EM (1996) Neuroglia and pioneer neurons express UNC-6 to provide global and local netrin cues for guiding migrations in *C. elegans* Neuron 16:35–46

Wadsworth WG (2006) UNC-6/netrin and SLT-1/slit guidance cues orient axon outgrowth mediated by MIG-10/RIAM/lamellipodin. Curr. Biol. 16:845–853

Wang X, Zhou F, Lv S, Yi P, Zhu Z, Yang Y, Feng G, Li W, Ou G (2013) Transmembrane protein MIG-13 links the Wnt signaling and Hox genes to the cell polarity in neuronal migration Proc Natl Acad Sci U S A 110(27):11175–80

Whangbo J, Kenyon C (1999) A Wnt signaling system that specifies two patterns of cell migration in *C. elegans*. Mol Cell 4:851–858

Wightman B, Clark SG, Taskar AM, Forrester WC, Maricq AV, Bargmann CI, Garriga G (1996) The *C. elegans* gene *vab-8* guides posteriorly directed axon outgrowth and cell migration. Development 122:671–682

Wolf FW, Hung MS, Wightman B, Way J, Garriga G (1998) *vab-8* is a key regulator of posteriorly directed migrations in *C. elegans* and encodes a novel protein with kinesin motor similarity Neuron 20(4):655–66

Yee C, Florica R, Fillingham J, Killeen MT (2014) ENU-3 functions in an UNC-6/Netrin dependent pathway parallel to UNC-40/DCC/Frazzled for outgrowth and guidance of the touch receptor neurons in *C. elegans*. Dev Dyn. 2013 Oct 7

Yee CS, Sybingco SS, Serdetchania V, Kholkina G, Bueno de Mesquita M, Naqvi, Z, Park SH, Lam K, Killeen MT (2011) ENU-3 is a novel motor axon outgrowth and guidance protein in *C. elegans*. Dev Biol 352:243–253

Yeo SY, Little MH, Yamada T, Miyashita T, Halloran MC, Kuwada JY, Huh TL, Okamoto H (2001) Overexpression of a slit homologue impairs convergent extension of the mesoderm and causes cyclopia in embryonic zebrafish. Dev Biol 230:1–17

Yoshikawa S, McKinnon RD, Kokel M, Thomas JB (2003) Wnt-mediated axon guidance via the *Drosophila* derailed receptor. Nature 422:583–588

Yu TW, Hao JC, Lim W, Tessier-Lavigne M, Bargmann CI (2002) Shared receptors in axon guidance: SAX-3/Robo signals via UNC-34/Enabled and a Netrin-independent UNC-40/DCC function. Nat Neurosci. 5(11):1147–1154

Yuan W, Zhou L, Chen JH, Wu JY, Rao Y, Ornitz DM (1999) The mouse SLIT family: secreted ligands for ROBO expressed in patterns that suggest a role in morphogenesis and axon guidance. Dev. Biol. 212:290–306

Zallen JA, Yi BA, Bargmann CI (1998) The conserved immunoglobulin superfamily member SAX-3/Robo directs multiple aspects of axon guidance in *C. elegans* Cell 92(2):217–227

Zallen JA, Kirch SA, Bargmann CI (1999) Genes required for axon pathfinding and extension in the *C. elegans* nerve ring. Development 126(16):3679–3692

Zinovyeva AY, Yamamoto Y, Sawa H, Forrester WC (2008) Complex network of Wnt signaling regulates neuronal migrations during *Caenorhabditis elegans* development. Genetics 179:1357–1371

Zou Y, Stoeckli E, Chen H, Tessier-Lavigne M (1999) Squeezing axons out of the gray matter: a role for slit and semaphorin proteins from midline and ventral spinal cord Cell 102(3):363–375

Chapter 3
Regulatory Mechanisms of Guidance Molecules During Growth Cone Migration and Axon Outgrowth

Abstract The most enigmatic aspect of metazoan nervous system development is that just a handful of cues and receptors steer numerous growth cones in an exceedingly complex environment to generate the final functional connectome. Studies in the past decades in both vertebrates and invertebrates have supported toward the existence of regulatory mechanisms of guidance molecules, transcriptionally, posttranscriptionally, and posttranslationally. This chapter focuses on posttranslational regulatory mechanisms including modular cues and receptors, localization, and proteolysis.

Keywords UNC-6/Netrin · UNC-40/DCC · UNC-5/UNC-5 · SLT-1/Slit · Receptor regulation

3.1 Modular Cues and Receptors

Structural and functional analysis of has revealed that most of the cues and receptors implicated in growth cone (GC) migration are modular; implicating that they are composed of different domains, where each domain interacts with specific set of signaling molecules, and each signaling module is associated with a specific guidance response. This section discusses about the modular nature of cues and receptors, involved in GC migration.

3.1.1 UNC-6/Netrin and SLT-1/Slit

UNC-6/Netrin, a 591-amino acid (aa) residue protein, functions as bifunctional cue, expressed in neuroglia and pioneer neurons, and its expression is both spatially and temporally regulated (Wadsworth et al. 1996). It is composed of an N-terminus (domains: VI, V-1, V-2, V-3) homologous to the N-termini of laminin B2 subunit, and a C-terminus (domain C), found in proteins including frizzled-related proteins, the

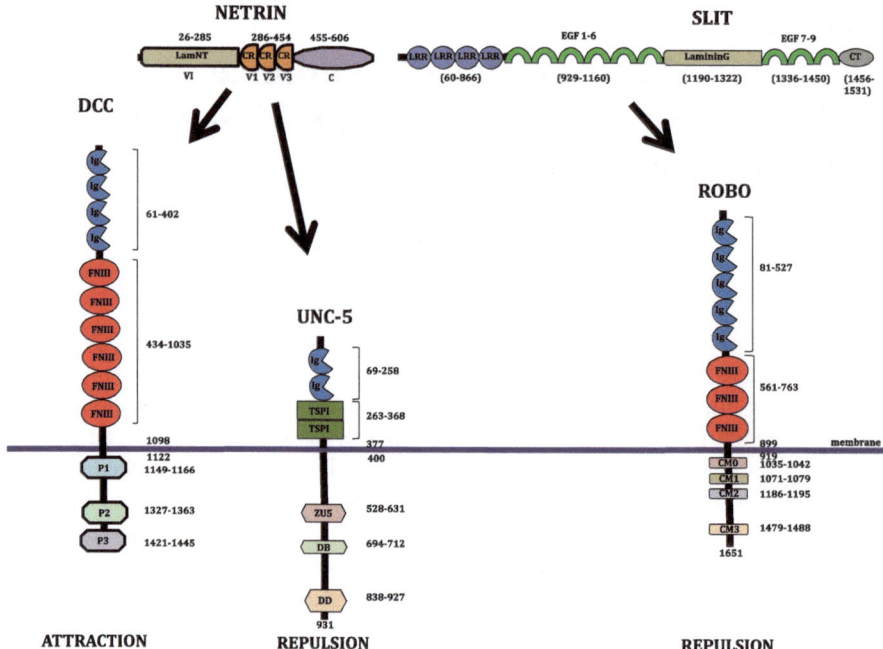

Fig. 3.1 Diagrammatic representation of modular structure of vertebrate cell migration and axon guidance molecules: structure of Netrin, a 606-aa residue protein, with a N-terminus VI, V-1, V-2, and V-3 domains and a C-terminus C-domain, functions as an attractive cue signaling through DCC (deleted in colorectal cancer) (1445-aa) receptor with four immunoglobulin (*Ig*) domains, six fibronectin type III (*FNIII*) domains, comprising the extracellular region, a cytoplasmic region composed of conserved motifs P1, P2, and P3, and a transmembrane domain (*TMD*). Netrin functions as a repulsive cue signaling via UNC-5 (931-aa) receptor, composed of two Ig and two thrombospondin type 1 (*TSPI*) domain forming the extracellular region, a cytoplasmic region composed of zona occludens 1 (ZU5), DCC binding (DB), and death domain (*DD*), a transmembrane domain (TMD). Slit (1531-aa), a repulsive ligand with four leucine-rich repeats (*LRRs*), nine epidermal growth factor repeats (*EGF*), a lamininG homology region, and a C-terminus cysteine knot region acts through its Robo receptor, which is composed of five Ig repeats, three FNIII repeats comprising the extracellular region and four conserved motifs CM0/CC0, CM1/CC1, CM2/CC2, and CM3/CC3 comprising cytoplasmic domain. The transmembrane domain (*TMD*) of the receptors is indicated as the starting and ending amino acids (-*aa*) number at the membrane. Position of each domain in the protein is indicated by number of -aa in each domain

complement C345 protein family, tissue inhibitors of metalloproteinases (TIMPs), and type I C-proteinase enhancer proteins (PCOLCEs) (Fig. 3.1) (Ishii et al. 1992; Serafini et al. 1994; Leyns et al. 1997; Banyai and Patthy 1999). Vertebrate Netrin is 606-aa residues and structurally similar to UNC-6 (Fig. 3.2) (Serafini et al. 1994).

Structural and functional analysis indicated that UNC-6 functions as bifunctional cue by signaling through two modules composed of two receptors including UNC-5 and UNC-40. Expressing various UNC-6 domain deletions in mutants with selective *unc-6*, loss-of-function alleles, associated dorsal migrations with VI, V-2, and V-3 domains and ventral migrations with VI and V-3 domains, carrying

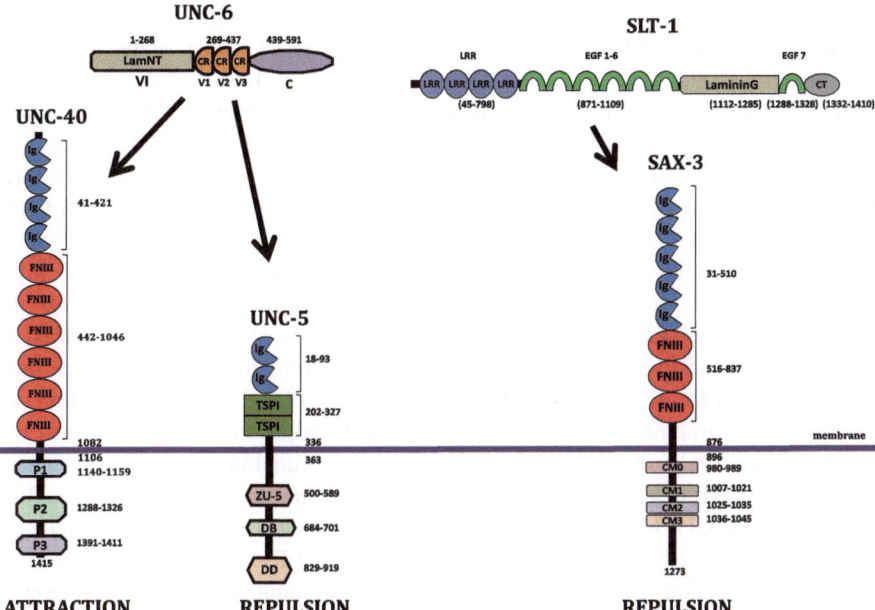

Fig. 3.2 Illustration of modular structure of *C. elegans* cell migration and axon guidance molecules: UNC-6, a 591-aa residue protein, is composed of a N-terminus VI, V-1, V-2, and V-3 domains and a C-terminus C domain, functions as an attractive cue signaling through UNC-40 (1415-aa) receptor with four immunoglobulin (*Ig*) domains, six fibronectin type III (*FNIII*) domains, comprising the extracellular region, a cytoplasmic region composed of conserved motifs P1, P2, and P3, and a transmembrane domain (*TMD*). UNC-6 functions as a repulsive cue signaling via UNC-5 (919-aa) receptor, composed of two Ig and two thrombospondin type 1 (*TSPI*) domain forming the extracellular region, a cytoplasmic region composed of zona occludens 1 (ZU5), DCC binding (DB), and death domain (*DD*). SLT-1, a repulsive ligand with four leucine-rich repeats (*LRRs*), seven epidermal growth factor repeats (*EGF*), a laminin-G homology region, and a C-terminus cysteine knot region, acts through its SAX-3 (1279-aa) receptor, which is composed of five Ig repeats, three FNIII repeats comprising the extracellular region and four conserved motifs CM0/CC0, CM1/CC1, CM2/CC2, and CM3/CC3 comprising cytoplasmic domain. The transmembrane domain (*TMD*) of the receptors is indicated as the starting and ending amino acids (*-aa*) number at the membrane. Position of each domain in the protein is indicated by number of -aa in each domain

out their responses by signaling via UNC-5 and UNC-40 receptors, respectively. *unc-6* loss-of-function and null alleles indicated that the domain VI produces defects in both direction- and tissue-specific guidance. Deletion of an eight aa motif within domain VI or deletion of the whole domain resulted in complete loss of *unc-6* activity, confirming its pivotal role in mediating various guidance responses (Wang and Wadsworth 2002).

Expressing *UNC-6 delta C* transgene (lacking domain C) in second messenger mutant background revealed that N-terminal domain induced branching and was sensitive to calmodulin kinase II (CaMKII) and diacylglycerol (DAG)-dependent signaling (Wadsworth et al. 1996; Lim et al. 1999; Wang and Wadsworth 2002).

Slit, a 1531-aa residue protein, conserved in both vertebrates and invertebrates mediates its repulsive function through Robo receptor. Sequence analysis revealed that *Drosophila* Slit has seven EGF repeats (Rothberg et al. 1988, 1990), whereas vertebrate Slits have nine EGF repeats (Holmes et al. 1998; Itoh et al. 1997; Nakayama et al. 1998; Brose et al. 1999; Li et al. 1999; Yuan et al. 1999), a laminin-like globular G domain (previously also known as the ALPS domain for agrin, laminin, perlecan, slit) followed by a cysteine-rich C-terminal region (Fig. 3.1). In *Caenorhabditis elegans*, SLT-1 is a 1410-aa residue protein, a Slit homologue, with essentially similar domain organization as it the *Drosophila* counterpart (Fig. 3.1) (Hao et al. 2001). Functional analysis of various mutant forms of vertebrate *slit* containing only the leucine-rich repeats (LRRs), showed that they function as chemorepellents for axons outgrowth. The LRRs can repel neurons migrating from the anterior subventricular zone (SVZa) to the olfactory bulb in brain slices isolated from neonatal rodents. The N-terminal fragments with either 890- or 1076-aa residues (including the LRR domains) are sufficient for both axon guidance and neuronal migration, a point mutation that encodes single amino acid change in the LRR-domain-reduced Slit signaling (Battye et al. 2001). Moreover, the LRRs were sufficient to bind to Ig domains of Robo (Chen et al. 2001). Similarly, in *Drosophila*, the LRRs mediate repellent signaling and D2 domain of LRR is required for in vitro binding to first two Ig domains of Robo (Liu et al. 2004; Howitt et al. 2004). Constructs-containing Slit D4 domain were used to study effect on chick retinal ganglion cell axons, supporting the role of D4 in Slit homodimerization and binding to heparin sulfate (Seiradake et al. 2009).

In *C. elegans*, phenotypic analysis of loss-of-function *slt-1* alleles on AVM (anterior ventral microtubule) ventral axon guidance implicated the second LRR with the repulsive function (Fig. 3.2) (Hao et al. 2001).

3.1.2 UNC-40/DCC/FRA, UNC-5/UNC-5, and SAX-3/Robo

Vertebrate, DCC (deleted in colorectal cancer) is a 1445-aa residue protein, characterized by the presence of an extracellular domain composed of a signal sequence at N-terminus, four immunoglobulin (Ig) C2 type repeats, six fibronectin (FN) type III repeats, a single transmembrane domain (TMD), and a cytoplasmic domain composed of three conserved motifs: P1, P2, and P3 (Fig. 3.2) (Hedrick et al. 1994; Vielmetter et al. 1994; Kolodziej et al. 1996). Structure and functional analysis of DCC associated the conserved motif P3 with Netrin-mediated attractive turning of stage 22 *Xenopus* neurons and also receptor multimerization (Stein et al. 2001). In flies, conserved P3 motif is required for GC attraction at the midline (Garbe et al. 2007). The *C. elegans* UNC-40 encodes a 1415-aa residue protein, with structure similar to vertebrate DCC (Fig. 3.1) (Chan et al. 1996; Gitai et al. 2003). Studies in *C. elegans* conferred conserved motifs P1 and P2 with a gain-of-function outgrowth response for UNC-40. Further genetic and biochemical analysis points toward UNC-34/Ena and CED-10 (Rac GTPase), to signal via the P1 and P2 motifs, respectively (Gitai et al. 2003).

The repulsive vertebrate receptor, UNC-5, a 931-aa residue protein, is composed of two Ig domains, two thrombospondin (TSP) type 1 domain, ZU5 (a portion homologous to zona occludens 1) (Itoh et al. 1997) region, DB domain (DCC binding) (Hong et al. 1999), and DD rehion (death domain) (Fig. 3.2) (Hofmann and Tschopp 1995). The *C. elegans* homolog, UNC-5, is a 919-aa residue protein; sequence analysis predicted domain organization and structure similar to vertebrate homologs (Fig. 3.1) (Leung-Hagesteijn et al. 1992; Killeen et al. 2002). In *C. elegans* based on locomotion assays, both the TSP-1 domains were partially functional for motor axon migrations. However, TSP-1(N) was preferentially required for DTC migrations. In addition to TSP-1(N) domain, rescue experiments carried out in *unc-5* mutants, using transgenes with deleted version of JM and ZU-5 regions, confirmed the requirement of these two regions for DTC migration and locomotion (Killeen et al. 2002).

Vertebrate Robo encodes for a single-pass transmembrane protein of around 1651-aa residues, composed of five Ig, three FNIII extracellular domains, and a cytoplasmic domain containing four conserved motifs (CMs/CCs) including CM0/CC0, CM1/CC1, CM2/CC2, and CM3/CC3 (Fig. 3.2) (Bashaw et al. 2000; Kidd et al. 1998). SAX-3/Robo, the *C. elegans* homolog, has a structure similar to vertebrate Robo; it has three isoforms with longest isoform predicted to encode 1273-aa residue protein (Fig. 3.1) (Zallen et al. 1998). Robo Ig domains function in *Drosophila* midline guidance, regulated by three Robo receptors, Robo/Robo1, Robo2, and Robo3, where Ig1 and Ig3 of Robo2 confer lateral positioning activity, and Ig2 confers promidline-crossing activity (Evans and Bashaw 2002). Additionally, studies with chimeric Robo receptors attributed Robo midline repulsion to CM1/CC1 and CM2/CC2 motif in its cytodomain (Spitzweck et al. 2010). Moreover, Robo interacts with srGAP3 (Slit–Robo GAP)/MEGAP with its CM2/CC2 and CM3/CC3 conserved motifs to regulate downstream signaling responses (Li et al. 2006).

It is evident from the experiments conducted in both vertebrates and invertebrates that modularity of guidance molecules regulates specific guidance responses either due to interaction with specific downstream signaling molecules or binding with receptors at the membrane.

3.2 Receptor Complexes

Besides the complex formation between the receptor and cues, several lines of evidence suggest that the receptor themselves form dimers and multimers to regulate the GC movement; complexes can be either homomeric or heteromeric.

Commissural axons in vertebrates experience simultaneously both attractive and repulsive cues at the midline. Studies have shown that it is the cytoplasmic portion of the Robo molecule that makes the guidance decisions at midline. In vitro studies have shown that Robo undergoes hemophilic interaction with Robo 2 and stimulates neurite outgrowth; this interaction requires the first Robo Ig domain (Hivert et al. 2002). In *Xenopus* spinal explants, it was found that activation of Robo silenced the attractive effect of Netrin-1 through direct binding of the Robo CC1/CM1 domain to the DCC conserved motif P3, causing silencing of the attractive effect of Netrin on neurons expressing both DCC and Robo (Stein and Tessier-Lavigne 2001). Co-immunoprecipitation assays further supported the asymmetric binding between Robo and DCC in a Slit-dependent manner; activation of Robo causes binding to DCC but reverse does not happen. Surprisingly, in contrast to vertebrate DCC, P3 does not mediate receptor self-association, and self-association is not sufficient to promote Fra-dependent attraction in *Drosophila* (Garbe et al. 2007). In transgenic *Drosophila*, phenotypic analysis using chimeric receptors including Fra–Robo (Fra's ectodomain and Robo's cytoplasmic domain) and Robo–Fra (Robo's ectodomain and Fra's cytoplasmic domain) specified the attractive versus repulsive function to the cytodomains of Robo and FRA/DCC (Bashaw and Goodman, 1999). The *C. elegans* SAX-3 receptor is also known to bind UNC-40, to form heterodimer, though the biological significance of this interaction is unknown (Yu et al. 2002).

Netrin are bifunctional cues attracting growth cones through their receptors DCC and repelling through UNC-5 but genetic evidence also suggests involvement of the DCC in some cases of repulsion (Chan et al. 1996; Leonardo et al. 1997). In vitro studies in *Xenopus* spinal axons to Netrin-1 suggested that attraction is converted to repulsion by UNC-5 expression, and repulsion requires DCC P1-conserved motif and UNC-5 DB domain. The cytoplasmic domains of UNC-5 and DCC were shown to interact physically, thereby converting DCC-mediated attraction to UNC-5/DCC-mediated repulsion in Netrin-1-dependent manner (Hong et al. 1999). Furthermore, loss of function of UNC-40 in worms also impairs migrations away from UNC-6/Netrin source (Hedgecock et al. 1990; McIntire et al. 1992; Colavita and Culotti 1998). *C. elegans*, DTC migration occurs in *unc-5-*, *unc-6-*, and *unc-40*-dependent manner and mutations in *unc-40* completely eliminate all dorsal guidance of hypomorphic *unc-5* allele, suggesting that DTC migration signaling results from the formation of a variety of oligomeric receptor complexes between UNC-40 and UNC-5 (Merz et al. 2001). Genetic and in vitro binding studies in *C. elegans* showed that UNC-40/DCC potentiates SAX-3/Robo signaling via a mechanism involving the direct binding of the two receptors through their cytoplasmic domains (Yu et al. 2002). SAX-3 receptor has been also proposed to bind EVA-1, a co-receptor for SLT-1 during ventral process guidance of AVM neuron; the binding between EVA-1 and SAX-3 has been confirmed by binding assays (Fujisawa et al. 2007). Based on both biochemical and genetic analyses, it is confirmed that guidance decisions are regulated by heteromeric receptor complex formation including UNC-40 binding to SAX-3 and UNC-5 in repulsive signaling and SAX-3 binding to UNC-40 and EVA-1 again in repulsive signaling decisions in worms.

3.3 Regulation of the Receptors by Phosphorylation

Phosphorylation and dephosphorylation act as molecular switches by either activating or deactivating proteins; in the former state, proteins can further stimulate downstream signaling pathways. Several families of tyrosine kinases and phosphatases have been identified as critical players in regulating transduction of information from specific guidance molecules (Yu and Bargmann 2001).

Focal adhesion kinase (FAK) and sarcoma (Src) kinases based on genetic and cell biological analysis are implicated in axon guidance and cell migration (Knoll and Drescher 2004), positively regulating each other (Tatosyan and Mizenina 2000). Both have been associated with Netrin signal transduction; in vitro studies have confirmed tyrosine phosphorylation of both DCC (Y1420) and UNC-5 in DCC-dependent manner (Li et al. 2004). Site-directed mutagenesis of Y1420 in DCC, reduced attractive GC extension of spinal neuron in response to Netrin (Li et al. 2004; Liu et al. 2004; Ren et al. 2004; Meriane et al. 2004).

In *C. elegans*, SRC-1 kinase phosphorylates tyrosine residue in UNC-5 receptor. In vitro binding assays in HEK292 cells confirmed SRC-1 interaction with UNC-5 through its SH2 domain. In vivo, *SRC-1* RNAi exhibited DTC migration defects, similar to *unc-5* mutants, and it suppressed the defects caused by UNC-5 overexpression, placing it downstream of UNC-5 receptor signaling (Lee et al. 2005).

Abelson tyrosine kinase (Abl), a cytoplasmic kinase and an oncogene, has been associated with axon extension (Lanier and Gertler 2000). Evidence supported the role of Abl in midline guidance by binding to conserved motif CC3/CM3 of Robo and phosphorylating tyrosine-1073 in CC1/CM1 motif, leading to attenuation of Slit response, and resulting in the absence of commissures due to constitutively active Abl (Bashaw et al. 2001). Moreover, biochemical assays in S2 cells show that Abl also complexes with and phosphorylates TRIO/UNC-73 and FRA/UNC-40. Significance of this interaction was confirmed by phenotypic analysis of *fra abl* and *fra trio* double mutants with loss of commissures in the *Drosophila* CNS (Forsthoefel et al. 2005).

3.4   Regulatory Mechanisms of Receptor Levels at the Membrane

The first step of GC extension and guidance involves the perception of signal followed by regulation of surface level of receptors and asymmetric localization of the guidance receptors. The following section discusses work done in vertebrates and invertebrates, providing insight into mechanisms of receptor regulation.

3.5   Trafficking of Cues and Receptors

Netrin/UNC-6 mediates its chemoattractant function through the DCC/UNC-40 receptor family members, promoting midline crossing of commissural axons. In embryonic rat commissural axons, surface level of DCC increases in Netrin-dependent manner; an effect further enhanced by protein kinase A (PKA or cAMP-dependent protein kinase A) activation. PKA probably regulates exocytosis of DCC, leading to Netrin-dependent increases in both surface expression of DCC and axon outgrowth (Bouchard et al. 2004). Interestingly, in cultured *Xenopus* neurons, Netrin did not directly influence PKA activity, suggesting that in rat commissural neurons, other signals are required to activate PKA. Studies indicate that Netrin-dependent inhibition of Rho activity might also contribute to DCC mobilization (Moore et al. 2008). Rho GTPases are known to be activated downstream of the receptors, but here the Rho GTPases may also have an upstream or feedback role in regulating the surface levels of guidance receptors by regulating actin dynamics, offering new insight that may help to explain how changing cyclic AMP (cAMP), PKA, and Rho activity promotes Netrin-mediated chemoattraction.

In *Drosophila*, Comm regulates the Robo level in the commissural axons by recruiting it to endosomes. Recent studies led to identification of Ras-related protein in brain guanine nucleotide dissociation inhibitor (RabGDI), to regulate Robo1 surface level in commissural axons. RabGDI is a regulator of Rab GTPases, which regulate many steps of membrane trafficking, including vesicle formation, vesicle movement along actin and tubulin networks, and membrane fusion (Stenmark and Olkkonen 2001). RabGDI is required for recycling of hydrolyzed RabGDP to RabGTP. Co-expression of RabGDI and Robo1 in COS cells enhanced the surface level of Robo1. Furthermore, localization of Robo1 in Rab11 vesicles, a marker for vesicles ready to insert cargo into the membrane, and the absence in Rab7 positive vesicles, a marker for the late endosomes and lysosomes, supported the hypothesis that RabGDI positively regulates Robo1 surface levels by exocytosis (Philipp et al. 2012).

In *C. elegans*, UNC-6 is expressed in glia, neurons, and muscle cells (Wadsworth et al. 1996; Asakura et al. 2007). A reverse genetic approach identified three genes involved in membrane trafficking of UNC-6 including *unc-14*, *unc-51*, and *unc-104*. Functional characterization of UNC-14, UNC-51, and UNC-104 homologs in other organisms predicted their importance in vesicle trafficking, including early endosome functions (McIntire et al. 1992; Tomoda et al. 2004; Sakamoto et al. 2005; Ogura and Goshima 2006; Toda et al. 2008). UNC-51, a serine/threonine kinase, binds UNC-14, a RUN domain protein (Ogura et al. 1994, 1997; Lai and Garriga 2004), whereas UNC-104 is a homolog of the kinesin motor protein KIF1A, which transports precursors of synaptic vesicles (SVs) and dense-core vesicles (DCVs) from neuronal cell bodies to synapses (Hall and Hedgecock 1991; Otsuka et al. 1991; Yonekawa et al. 1998; Zahn et al. 2004; Hirokawa and Noda 2008). Analysis of transgenic worms with *Venus::UNC-6* in *unc-51*, *unc-14*, and *unc-104* mutant background resulted in accumulation of UNC-6 primarily in the cell bodies, and very little UNC-6 was detected in axons and consecutively secreted (Ogura et al. 2012). A genetic interaction was also established between *unc-14*, *unc-51*, and *unc-104* with *unc-6* (Ogura and Goshima 2006; Asakura et al. 2010). Similarly, UNC-5 localization was regulated by UNC-51 and UNC-14 in neuronal cells (Ogura and Goshima 2006). In *unc-14* and *unc-51* mutants, UNC-5 accumulates at the neural cell bodies, and little UNC-5 was observed at axons in worms carrying *UNC-5::GFP* transgene (Ogura et al. 2012). Recent studies implicated UNC-51 in membrane localization of SAX-3 as observed in SAX-3::GFP transgenic worms, *unc-51* mutants exhibited *SAX-3::GFP* in the cell bodies of posterior deirid (PDE) neurons (Li et al. 2013). UNC-51 thereby regulates secretion of UNC-6 by regulating its localization and secretion in the axons and localization of UNC-5 and SAX-3 at the membrane.

Genetic approaches in *C. elegans* support the importance of regulating receptor trafficking in GC migration and guidance. Variable abnormal 8 (VAB-8), a kinesin-like protein, is required for posterior migration in worms. Animals expressing *Pmec-7::VAB-8L::GFP* in mechanosensory neurons showed ALM axon-rerouting phenotype, in wild-type worms; ALM extends an anterior process, VAB-8L expression resulted in polarity reversal, causing ALM to extend a posterior process. Genetic and biochemical analysis further indicated that VAB-8L functions in UNC-73/TRIO (Trio-family RacGEF)-dependent manner, stimulating small GTPase, MIG-2/RhoG for the localization of the SAX-3/Robo and UNC-40/DCC receptors to the membrane (Watari-Goshima et al. 2007; Levy-Strumpf and Culotti 2007). These studies support the role of Rho GTPases acting upstream of guidance receptors, which conventionally function downstream of receptors to regulate actin cytoskeleton. Furthermore, both VAB-8 and UNC-73 physically interact with UNC-51, suggesting VAB-8 and UNC-73 may cooperate with UNC-51 to regulate the localization or trafficking of these receptors. In *Drosophila*, ATG1/ UNC-51 kinase regulates the interaction between synaptic vesicles and motor complexes during transport. UNC-51 binds UNC-76, a kinesin heavy-chain (KHC) adaptor protein and loss of *UNC-51* or *UNC-76* function leads to severe axonal transport defects, segregating synaptic vesicles from the motor complexes and their accumulation along axons (Toda et al. 2008). Probably in *C. elegans*, UNC-51 uses a similar mechanism for exocytosis of receptors.

While VAB-8L positively regulates the level of guidance receptors at the membrane by activating UNC-73/Trio a RacGEF (Steven et al. 1998), to promote ALM-rerouting phenotype (Watari-Goshima et al. 2007), CRML-1 negatively regulates the level of SAX-3 receptor by inhibiting UNC-73/Trio supported by quantitation of GFP in ALM and its lineal sister BDU in worms with *SAX-3::GFP* transgene. Besides, CRML-1 and UNC-73 exist in a complex *in vivo* confirmed by co-immunoprecipitation assay. CRML-1 is a *C. elegans* homolog of CARMIL, an actin un-capping protein (Jung et al. 2001; Remmert et al. 2004). CRML-1 was identified in genetic screen as suppresser of *unc-34*. Genetic studies with *crml-1 unc-34* mutants placed it in a parallel pathway to *unc-34* as an inhibitor of cell migration defects of ALM and CAN neurons (Vanderzalm et al. 2009).

At the fly midline, Comm controls midline crossing by negatively regulating the Robo receptor in commissural axons, preventing commissural axons from responding to Slit the midline repellent. Comm mRNA is detected both in midline glia as well as for short time in commissural axons as they approach the midline. Transgenic expression of mutant forms of Comm and subcellular localization studies indicated that Comm regulates midline repulsion by recruiting Robo receptors to endosomes for degradation before they reach the cell surface (Keleman et al. 2002). Expression of Robo::GFP in sensory axons helped to visualize the anterograde axonal transport of Robo positive vesicles, which are undetectable when Comm is genetically introduced into these Robo::GFP-positive neurons

(Keleman et al. 2005). Additionally, a role for Nedd4, a ubiquitin ligase, has been implicated in Comm regulation of surface level of Robo, mutation in either the dNedd4 binding site or the ubiquitin acceptor sites in Comm disrupts its ability to regulate Robo (Myat et al. 2002). Interestingly, Comm homologs in mammals or *C. elegans* have not been identified pointing to existence of other proteins, which carry out the same function. Particularly, in mammals, Rig-1/Robo3, a divergent Robo family member, is required in precrossing commissural neurons to downregulate the sensitivity to midline Slit proteins. Genetic and in vitro analyses indicate that Rig-1 functions to repress Slit responsiveness, without affecting the level of Robo at the membrane (Sabatier et al. 2004).

3.6 Regulated Proteolytic Processing of Cues and Receptors

Proteolytic processing of guidance molecules can affect axon outgrowth and cell migration. Matrix metalloproteases or metalloprotinases (MMP) family of extracellular metalloproteases and a disintegrin and metalloproteases (ADAMs) are conserved throughout the animal kingdom and are known to be involved in several cellular processes including axon guidance (McFarlane 2003; Page-McCaw 2008). Several studies have implicated the ability of metalloproteases to cleave the ligands or receptors to inhibit or activate specific signaling pathways. ADAM-10 metalloproteases as well as matrix metalloproteases play important role in axon guidance in vivo in both invertebrate and vertebrate nervous systems (Chen et al. 2007; Hehr et al. 2005).

Kuzbanian, ADAM family transmembrane metalloprotease was identified in a genetic screen for defects in midline axonal guidance in *Drosophila* (Fambrough et al. 1996); mutations in *kuzbanian* (*kuz*) exhibited dose-dependent genetic interactions with *slit*. Ectopic midline crossing of ipsilateral interneurons was observed in *kuz* zygotic mutant embryos and in embryos where both *slit* and *kuz* function were partially reduced, supporting the hypothesis that Kuz may be a positive regulator of Slit–Robo signaling (Schimmelpfeng et al. 2001). Moreover, antibody staining for Robo1 in *kuz* mutant revealed that the midline phenotype is accompanied by a failure to exclude Robo1 protein expression from the midline-crossing portions of axons, suggesting that *kuz* activity may be necessary for exclusion of Robo1 from midline portion of axons. Together, the above experiments suggest that Kuz may regulate guidance by regulating the cleavage of Robo.

Similarly, effect on receptor expression in the context of metalloprotease-dependent ectodomain shedding of DCC was observed in in vitro experiments, where chemical inhibition of metalloprotease activity enhanced DCC receptor expression at the membrane and promoted Netrin-mediated axon growth, suggesting that proteolytic cleavage regulates level of receptors at the plasma membrane (Galko and Tessier-Lavigne 2000).

ADAM10 has also been linked with ephrin-A2 signaling via its ligand Eph and EphrinB/EphB (Hattori et al. 2000; Lin et al. 2008). In *C. elegans*, *unc-71* corresponding to *adm-1* is one of the four ADAM genes (Podbilewicz 1996) associated with axon guidance of motoneurons and migration of SMs, acting with UNC-53, a scaffolding protein (Marcus-Gueret et al. 2012), though the mechanism of UNC-71 action is still not clear.

Ligand-stimulated ubiquitination of receptors or associated proteins leads to the endocytosis of the receptors, followed by targeting to the proteasome for degradation. The ubiquitination of DCC/UNC-40/FRA was demonstrated in a heterologous system (Hu et al. 1997); ubiquitin and DCC were overexpressed in HEK cells resulting in ubiquitination and degradation of DCC in a Netrin-1-dependent manner. This process required ubiquitin ligase, *seven in absentia homologs* (*siahs*). Biochemical and genetic interaction between DCC and Siah was also established in *Drosophila* (Hu et al. 1997). Moreover, studies in *Xenopus* retinal ganglion cells in culture showed that the Netrin-induced GC turning is blocked by inhibition of proteasome function and that Netrin stimulates the production of unidentified ubiquitin conjugates in these GCs (Campbell and Holt 2001).

In addition to the positive regulatory mechanism of the receptors, the trafficking of SAX-3 and UNC-5 was shown to be negatively regulated, affecting axon outgrowth. A genetic screen in *C. elegans* for genes that could modulate UNC-6 signaling identified *rpm-1* mutation, encoding E3 ubiquitin ligase, a member of the conserved Pam/Highwire/RPM protein family that plays important roles in presynaptic differentiation (Li et al. 2008). In genetic backgrounds where *sax-3* and *unc-5* function is partially reduced, *rpm-1* mutation resulted in specific axon overgrowth and branching phenotypes. Moreover, SAX-3 and UNC-5 proteins exhibited increased expression levels and altered localization in *rpm-1* mutants. Genetic analysis indicated the role of RPM-1 in regulating SAX-3 and UNC-5 is dependent on GLO-4, a RAB (Ras-related in brain) guanine nucleotide exchange factor (GEF), implicated in vesicle trafficking (Grill et al. 2007), indicating an important role for protein trafficking in axon growth regulation (Li et al. 2008). CLEC-38, a transmembrane protein with C-type lectin-like domains (CTLDs), an axon guidance regulator, regulates the expression UNC-40 receptor observed in worms with UNC-40::GFP transgene, and probably, it requires RPM-1 function for this process supported by the genetic data, indicating a role for CLEC-38 in UNC-40-dependent GC migration (Kulkarni et al. 2008; Li et al. 2008). Tripartite motif (TRIM) family proteins are ring finger domain-containing, multidomain proteins. *C. elegans* homolog, TRIM-9, has been implicated in UNC-6 UNC-40 attractive guidance pathway in HSN and AVM neurons. Asymmetric localization of MIG-10, a scaffolding protein and a downstream effector of UNC-40, was abolished in *trim-9* mutants. In vitro experiments showed that TRIM-9 has E3 ubiquitin ligase activity, which is important for its in vivo function (Song 2011). Ubiquitin ligases are enzymes polyubiquinating and marking proteins for degradation by proteasomes. Monoubiquination on the other hand results in altered function of a protein either by changing its cellular location or interaction with other proteins.

References

Asakura T, Ogura K, Goshima Y (2007) UNC-6 expression by the vulval precursor cells of *Caenorhabditis elegans* is required for the complex axon guidance of the HSN neurons. Dev Biol 304:800–810

Asakura T, Waga N, Ogura K, Goshima Y (2010) Genes required for cellular UNC-6/netrin localization in *Caenorhabditis elegans*. Genetics 185:573–585

Banyai L, Patthy L (1999) The NTR module: domains of netrins, secreted frizzled related proteins, and type I procollagen C-proteinase enhancer protein are homologous with tissue inhibitors of metalloproteases. Protein Sci 8:1636–1642

Bashaw GJ, Goodman CS (1999) Chimeric axon guidance receptors: the cytoplasmic domains of Slit and Netrin receptors specify attraction versus repulsion. Cell 97:917–926

Bashaw GJ, Kidd T, Murray D, Pawson T, Goodman CS (2000) Repulsive axon guidance: Abelson and enabled play opposing roles downstream of the roundabout receptor. Cell 101(7):703–715

Bashaw GJ, Hu H, Nobes CD, Goodman CS (2001) A novel Dbl family RhoGEF promotes Rho-dependent axon attraction to the central nervous system midline in *Drosophila* and overcomes Robo repulsion. J Cell Biol 155(7):1117–1122

Battye R, Stevens A, Perry RL, Jacobs JR (2001) Repellent signaling by Slit requires the Leucine-Rich repeats. J Neurosci 21(12):4290–4298

Bouchard JF, Moore SW, Tritsch NX, Roux PP, Shekarabi M, Barker PA, Kennedy TE (2004) Protein kinase a activation promotes plasma membrane insertion of DCC from an intracellular pool: a novel mechanism regulating commissural axon extension. J Neurosci 24(12):3040–3050

Brose K, Bland KS, Wang KH, Arnott D, Henzel W, Goodman CS, Tessier-Lavigne M, Kidd T (1999) Slit proteins bind Robo receptors and have an evolutionarily conserved role in repulsive axon guidance. Cell 96(6):795–806

Campbell DS, Holt CE (2001) Chemotropic responses of retinal growth cones mediated by rapid local protein synthesis and degradation. Neuron 32:1013–1026

Chan SS, Zheng H, Su MW, Wild R, Killeen MT, Hedgecock EM, Culotti JG (1996) UNC-40, a *C. elegans* homolog of DCC (deleted in colorectal cancer), is required in motile cells responding to UNC-6 Netrin cues. Cell 87:187–195

Chen JH, Wen L, Dupuis S, Wu JY, Rao Y (2001) The N-terminal leucine-rich regions in Slit are sufficient to repel olfactory bulb axons and subventricular zone neurons. J Neurosci 21(5):1548–1556

Chen YY, Hehr CL, Atkinson-Leadbeater K, Hocking JC, McFarlane S (2007) Targeting of retinal axons requires the metalloproteinase ADAM10. J Neurosci 27:8448–8456

Colavita A, Culotti JG (1998) Suppressors of ectopic UNC-5 growth cone steering identify eight genes involved in axon guidance in *Caenorhabditis elegans*. Dev Biol 194:72–85

Evans TA, Bashaw GJ (2002) Functional diversity of Robo receptor immunoglobulin domains promotes distinct axon guidance decisions. Curr Biol 20(6):567–572

Fambrough D, Pan D, Rubin GM, Goodman CS (1996) The cell surface metalloprotease/disintegrin Kuzbanian is required for axonal extension in *Drosophila*. Proc Natl Acad Sci USA 93:13233–13238

Forsthoefel DJ, Liebl EC, Kolodziej PA, Seeger MA (2005) The Abelson tyrosine kinase, the Trio GEF and enabled interact with the Netrin receptor Frazzled in *Drosophila*. Development 132(8):1983–1994

Fujisawa K, Wrana JL, Culotti JG (2007) The Slit receptor EVA-1 coactivates a SAX-3/ Robo mediated guidance signal in *C. elegans*. Science 317:1934–1938 (Erratum in: Science 318:570)

Galko MJ, Tessier-Lavigne M (2000) Function of an axonal chemoattractant modulated by metalloprotease activity. Science 289:1365–1367

Garbe DS, O'Donnell M, Bashaw GJ (2007) Cytoplasmic domain requirements for Frazzled-mediated attractive axon turning at the *Drosophila* midline. Development 134(24):4325–4334

Gitai Z, Yu TW, Lundquist EA, Tessier-Lavigne M, Bargmann CI (2003) The netrin receptor UNC-40/DCC stimulates axon attraction and outgrowth through enabled and, in parallel, Rac and UNC-115/AbLIM. Neuron 9, 37(1):53–65

Grill B, Bienvenut WV, Brown HM, Ackley BD, Quadroni M, Jin Y (2007) *C. elegans* RPM-1 regulates axon termination and synaptogenesis through the Rab GEF GLO-4 and the Rab GTPase GLO-1. Neuron 55:587–601

Hall DH, Hedgecock EM (1991) Kinesin-related gene *unc-104* is required for axonal transport of synaptic vesicles in *C. elegans*. Cell 65:837–847

Hao JC, Yu TW, Fujisawa K, Culotti JG, Gengyo-Ando K, Mitani S, Moulder G, Barstead R, Tessier-Lavigne M, Bargmann CI (2001) *C. elegans* Slit acts in midline, dorsal–ventral, and anterior–posterior guidance via the SAX-3/Robo receptor. Neuron 32:25–38

Hattori M, Osterfield M, Flanagan JG (2000) Regulated cleavage of a contact-mediated axon repellent. Science 289:1360–1365

Hedgecock EM, Culotti JG, Hall DH (1990) The *unc-5*, *unc-6*, and *unc-40* genes guide circumferential migrations of pioneer axons and mesodermal cells on the epidermis in *C. elegans*. Neuron 4:61–85

Hedrick L, Cho KR, Fearon ER, Wu T-C, Kinzler KW, Vogelstein B (1994) The DCC gene product in cellular differentiation and colorectal tumorigenesis. Genes Dev 8:1174–1183

Hehr CL, Hocking JC, McFarlane S (2005) Matrix metalloproteinases are required for retinal ganglion cell axon guidance at select decision points. Development 132:3371–3379

Hirokawa N, Noda Y (2008) Intracellular transport and kinesin superfamily proteins, KIFs: structure, function, and dynamics. Physiol Rev 88:1089–1118

Hivert B, Liu Z, Chuang CY, Doherty P, Sundaresan V (2002) Robo1 and Robo2 are homophilic binding molecules that promote axonal growth. Mol Cell Neurosci 21(4):534–545

Hofmann K, Tschopp J (1995) The death domain motif found in Fas (Apo-1) and TNF receptor is present in proteins involved in apoptosis and axonal guidance. FEBS Lett 371(3):321–323

Holmes GP, Negus K, Burridge L, Raman S, Algar E, Yamada T, Little MH (1998) Distinct but overlapping expression patterns of two vertebrate slit homologs implies functional roles in CNS development and organogenesis. Mech Dev 79:57–72

Hong K, Hinck L, Nishiyama M, Poo MM, Tessier-Lavigne M, Stein E (1999) A ligand-gated association between cytoplasmic domains of UNC5 and DCC family receptors converts netrin—induced growth cone attraction to repulsion. Cell 97:927–941

Howitt JA, Clout NJ, Hohenester E (2004) Binding site for Robo receptors revealed by dissection of the leucine-rich repeat region of Slit. EMBO J 23:4406–4412

Hu G, Zhang S, Vidal M, Baer JL, Xu T, Fearon ER (1997) Mammalian homologs of seven in absentia regulate DCC via the ubiquitin-proteasome pathway. Genes Dev 11:2701–2714

Ishii N, Wadsworth WG, Stern BD, Culotti JG, Hedgecock EM (1992) UNC-6, a laminin-related protein, guides cell and pioneer axon migrations in *C. elegans*. Neuron 9:873–881

Itoh A, Miyabayashi T, Ohno M, Sakano S (1997) Cloning and expressions of three mammalian homologues of *Drosophila* slit suggest possible roles for slit in the formation and maintenance of the nervous system. Brain Res Mol Brain Res 62:175–186

Jung G, Remmert K, Wu X, Volosky JM, Hammer JA 3rd (2001) The *Dictyostelium* CARMIL protein links capping protein and the Arp2/3 complex to type I myosins through their SH3 domains. J Cell Biol 153:1479–1497

Keleman K, Rajagopalan S, Cleppien D, Teis D, Paiha K, Huber LA, Technau GM, Dickson BJ (2002) Comm sorts robo to control axon guidance at the *Drosophila* midline. Cell 110:415–427

Keleman K, Ribeiro C, Dikson BJ (2005) Comm function in commissural axon guidance: cell-autonomous sorting of Robo in vivo. Nat Neurosci 8:156–163

Kidd T, Brose K, Mitchell KJ, Fetter RD, Tessier-Lavigne M, Goodman CS, Tear G (1998) Roundabout controls axon crossing of the CNS midline and defines a novel subfamily of evolutionarily conserved guidance receptors. Cell 92(2):205–215

Killeen MT, Tong J, Krizus A, Steven R, Scott I, Pawson T, Culotti J (2002) UNC-5 function requires phosphorylation of cytoplasmic tyrosine 482, but its UNC-40-independent functions also require a region between the ZU-5 and death domains. Dev Biol 251:348–366

Knoll B, Drescher U (2004) Src family kinases are involved in Eph receptor-mediated retinal axon guidance.J.Neurosci 24:6248-6257

Kolodziej PA, Timpe LC, Mitchell KJ, Fried SR, Goodman CS, Jan LY, Jan YN (1996) Frazzled encodes a *Drosophila* member of the DCC immunoglobulin subfamily and is required for CNS and motor axon guidance. Cell 87:197–204

Kulkarni G, Li H, Wadsworth WG (2008) CLEC-38, a transmembrane protein with C-Type lectin-like domains, negatively regulates UNC-40-mediated axon outgrowth and promotes presynaptic development in *Caenorhabditis elegans*. J Neurosci 28:4541–4550

Lai T, Garriga G (2004) The conserved kinase UNC-51 acts with VAB-8 and UNC-14 to regulate axon outgrowth in *C. elegans*. Development 131(23):5991–6000

Lanier LM, Gertler FB (2000) Actin cytoskeleton: thinking globally, actin' locally. Curr Biol 10(18):655–657

Lee J, Li W, Guan KL (2005) SRC-1 mediates UNC-5 signaling in *Caenorhabditis elegans*. Mol Cell Biol 25(15):6485–6495

Leonardo ED, Hinck L, Masu M, Keino-Masu K, Ackerman SL,Tessier-Lavigne M (1997) Vertebrate homologues of C.elegans UNC-5 are candidate netrin receptors. Nature 386(6627):833–838

Leung-Hagesteijn C, Spence AM, Stern BD, Zhou Y, Su MW, Hedgecock EM, Culotti JG (1992) UNC-5, a transmembrane protein with immunoglobulin and thrombospondin type 1 domains, guides cell and pioneer axon migrations in *C. elegans*. Cell 71(2):289–299

Levy-Strumpf N, Culotti JG (2007) VAB-8, UNC-73 and MIG-2 regulate axon polarity and cell migration functions of UNC-40 in *C. elegans*. Nat Neurosci 10:161–168

Leyns L, Bouwmeester T, Kim SH, Piccolo S, De Robertis EM (1997) Frzb-1 is a secreted antagonist of Wnt signaling expressed in the Spemann organizer. Cell 88:747–756

Li HS, Chen JH, Wu W, Fagaly T, Zhou L, Yuan W, Dupuis S, Jiang ZH, Nash W, Gick C, Ornitz DM, Wu JY, Rao Y (1999) Vertebrate *slit*, a secreted ligand for the transmembrane protein roundabout, is a repellent for olfactory bulb axons. Cell 96:807–818

Li W, Lee J, Vikis HG, Lee SH, Liu G, Aurandt J, Shen TL, Fearon ER, Guan JL, Han M, Rao Y, Hong K, Guan KL (2004) Activation of FAK and Src are receptor-proximal events required for netrin signaling. Nat Neurosci 7(11):1213–1221

Li H, Kulkarni G, Wadsworth WG (2008) RPM-1, a *Caenorhabditis elegans* protein that functions in presynaptic differentiation, negatively regulates axon outgrowth by controlling SAX-3/robo and UNC-5/UNC5 activity. J Neurosci 28:3595–3603

Li X, Chen Y, Liu Y, Gao J, Gao F, Bartlam M, Wu JY, Rao Z (2006) Structural basis of Robo proline-rich motif recognition by the srGAP1 Src homology 3 domain in the Slit-Robo signaling pathway. J Biol Chem 281(38):28430–28437

Li J, Pu P, Le W (2013) The SAX-3 receptor stimulates axon outgrowth and the signal sequence and transmembrane domain are critical for SAX-3 membrane localization in the PDE neuron of *C. elegans*. PLoS One 8(6):12

Lim YS, Mallapur S, Kao G, Ren XC, Wadsworth WG (1999) Netrin UNC-6 and the regulation of branching and extension of motoneuron axons from the ventral nerve cord of *Caenorhabditis elegans*. J Neurosci 19:7048–7056

Lin KT, Sloniowski S, Ethell DW, Ethell IM (2008) Ephrin-B2 induced cleavage of EphB2 receptor is mediated by matrix metalloproteinases to trigger cell repulsion. J Biol. Chem. 283:28969–28979

Liu G, Beggs H, Jurgensen C, Park HT, Tang H, Gorski J, Jones KR, Reichardt LF, Wu J, Rao Y (2004) Netrin requires focal adhesion kinase and Src family kinases for axon outgrowth and attraction. Nat Neurosci 7:1222–1232

Marcus-Gueret N, Schmidt KL, Stringham EG (2012) Distinct cell guidance pathways controlled by the Rac and Rho GEF domains of UNC-73/TRIO in *Caenorhabditis elegans*. Genetics 190:129–142

McFarlane S (2003) Metalloproteases: carving out a role in axon guidance. Neuron 37(4):559–562

McIntire SL, Garriga G, White J, Jacobson D, Horvitz HR (1992) Genes necessary for directed axonal elongation or fasciculation in *C. elegans*. Neuron 8:307–322

Meriane M, Tcherkezian J, Webber CA, Danek EI, Triki I, McFarlane S, Bloch-Gallego E, Lamarche-Vane N (2004) Phosphorylation of DCC by Fyn mediates Netrin-1 signaling in growth cone guidance. J Cell Biol 167:687–698

Merz DC, Zheng H, Killeen MT, Krizus A, Culotti JG (2001) Multiple signaling mechanisms of the UNC-6/netrin receptors UNC-5 and UNC-40/DCC in vivo. Genetics 158(3):1071–1080

Moore SW, Correia JP, Lai Wing Sun K, Pool M, Fournier AE, Kennedy TE (2008) Rho inhibition recruits DCC to the neuronal plasma membrane and enhances axon chemoattraction to *netrin 1*. Development 135:2855–2864

Myat A, Henry P, McCabe V, Flintoft L, Rotin D, Tear G (2002) *Drosophila* Nedd4, a ubiquitin ligase, is recruited by commissureless to control cell surface levels of the roundabout receptor. Neuron 35:447–459

Nakayama M, Nakajima D, Nagase T, Nomura N, Seki N, Ohara O (1998) Identification of high-molecular-weight proteins with multiple EGF-like motifs by motif-trap screening. Genomics 51:27–34

Ogura K, Goshima Y (2006) The autophagy-related kinase UNC-51 and its binding partner UNC-14 regulate the subcellular localization of the netrin receptor UNC-5 in *Caenorhabditis elegans*. Development 133:3441–3450

Ogura K, Shirakawa M, Barnes TM, Hekimi S, Ohshima Y (1997) The UNC-14 protein required for axonal elongation and guidance in Caenorhabditis elegans interacts with the serine/threonine kinase UNC-51. Genes Dev 11(14):1801–1811

Ogura K, Wicky C, Magnenat L, Tobler H, Mori I, Müller F, Ohshima Y (1994) Caenorhabditis elegans unc-51 gene required for axonal elongation encodes a novel serine/threonine kinase.

Ogura K, Asakura T, Goshima Y (2012) Localization mechanisms of the axon guidance molecule UNC-6/Netrin and its receptors, UNC-5 and UNC-40, in *Caenorhabditis elegans*. Dev Growth Differ 54:390–397

Otsuka AJ, Jeyaprakash A, Garcia-Anoveros J, Tang LZ, Fisk G, Hartshorne T, Franco R, Born T (1991) The *C. elegans unc-104* gene encodes a putative kinesin heavy chain-like protein. Neuron 6:113–122

Page-McCaw A (2008) Remodeling the model organism: matrix metalloproteinase functions in invertebrates. Semin Cell Dev Biol 19(1):14–23

Philipp M, Niederkofler V, Debrunner M, Alther T, Kunz B, Stoeckli ET (2012) RabGDI controls axonal midline crossing by regulating Robo1 surface expression. Neural Dev 9(7):36

Podbilewicz B (1996) ADM-1, a protein with metalloprotease- and disintegrin- like domains, is expressed in syncytial organs, sperm, and sheath cells of sensory organs in *Caenorhabditis elegans*. Mol Bio Cell 7:1877–1893

Remmert K, Olszewski TE, Bowers MB, Dimitrova M, Ginsburg A, Hammer JA 3rd (2004) CARMIL is a bona fide capping protein interactant. J Biol Chem 279:3068–3077

Ren XR, Ming GL, Xie Y, Hong Y, Sun DM, Zhao ZQ, Feng Z, Wang Q, Shim S, Chen ZF, Song HJ, Mei L, Xiong WC (2004) Focal adhesion kinase in netrin-1 signaling. Nat Neurosci 7(11):1204–1212

Rothberg JM, Hartley DA, Walther Z, Artavanis-Tsakonas S (1988) *Slit*: an EGF-homologous locus of *D. melanogaster* involved in the development of the embryonic central nervous system. Cell 55:1047–1059

Rothberg JM, Jacobs JR, Goodman CS, Artavanis-Tsakonas S (1990) *Slit*: an extracellular protein necessary for development of midline glia and commissural axon pathways contains both EGF and LRR domains. Genes Dev 4:2169–2187

Sabatier C, Plump AS, Ma Le, Brose K, Tamada A, Murakami F, Lee EV, Tessier-Lavigne M (2004) The divergent Robo family protein *rig-1*/Robo3 is a negative regulator of Slit responsiveness required for midline crossing by commissural axons. Cell 117:157–169

Sakamoto R, Byrd DT, Brown HM, Hisamoto N, Mat-sumoto K, Jin Y (2005) The *Caenorhabditis elegans* UNC-14 RUN domain protein binds to the *kinesin-1* and UNC-16 complex and regulates synaptic vesicle localization. Mol Biol Cell 16:483–496

Schimmelpfeng K, Gogel S, Klambt C (2001) The function of leak and kuzbanian during growth cone and cell migration. Mech Dev 106:25–36

Seiradake E, von Philipsborn AC, Henry M, Fritz M, Lortat-Jacob H, Jamin M, Hemrika W, Bastmeyer M, Cusack S, McCarthy AA (2009) Structure and functional relevance of the Slit2 homodimerization domain. EMBO Rep 10(7):736–741

Serafini T, Kennedy TE, Galko MJ, Mirzayan C, Jessell TM, Tessier-Lavigne M (1994) The Netrins define a family of axon outgrowth-promoting proteins homologous to *C. elegans*-6. Cell 78:409–424

Song S, Ge Q, Wang J, Chen H, Tang S, Bi J, Li X, Xie Q, Huang X (2011) TRIM-9 functions in the UNC-6/UNC-40 pathway to regulate ventral guidance. J Genet Genomics 38(1):1–11

Spitzweck B, Brankatschk M, Dickson BJ (2010) Distinct protein domains and expression patterns confer divergent axon guidance functions for *Drosophila* Robo receptors. Cell 140(3):409–420

Stein E, Tessier-Lavigne M (2001) Hierarchical organization of guidance receptors: silencing of netrin attraction by Slit through a Robo/DCC receptor complex. Science 291:1928–1938

Stein E, Zou Y, Poo M, Tessier-Lavigne M (2001) Binding of DCC by netrin-1 to mediate axon guidance independent of adenosine A2B receptor activation. Science 291:1976–1982

Stenmark H, Olkkonen VM (2001) The Rab GTPase family. Genome Biol 2(5):REVIEWS3007

Steven R, Kubiseski TJ, Zheng H, Kulkarni S, Mancillas J, Ruiz Morales A, Hogue CW, Pawson T, Culotti J (1998) UNC-73 activates the Rac GTPase and is required for cell and growth cone migrations in *C. elegans*. Cell 92:785–795

Tatosyan AG, Mizenina OA (2000) Kinases of the Src family: structure and functions. Biochemistry (Mosc) 65:49–58

Toda H, Mochizuki H, Flores R 3rd, Josowitz R, Krasieva TB, Lamorte VJ, Suzuki E, Gindhart JG, Furukubo-Tokunaga K, Tomoda T (2008) UNC-51/ATG1 kinase regulates axonal transport by mediating motor—cargo assembly. Genes Dev 22:3292–3307

Tomoda T, Kim JH, Zhan C, Hatten ME (2004) Role of Unc51.1 and its binding partners in CNS axon outgrowth. Genes Dev 18:541–558

Vanderzalm PJ, Pandey A, Hurwitz ME, Bloom L, Horvitz HR, Garriga G (2009) *C. elegans* CARMIL negatively regulates UNC-73/Trio function during neuronal development. Development 136(7):1201–1210

Vielmetter J, Kayyem JE, Roman JM, Dreyer WJ (1994) Neogenin, an avian cell surface protein expressed during terminal neuronal differentiation, is closely related to the human tumor suppressor molecule deleted in colorectal cancer. J Cell Biol 127(6):2009–2020

Wadsworth WG, Bhatt H, Hedgecock EM (1996) Neuroglia and pioneer neurons express UNC-6 to provide global and local netrin cues for guiding migrations in *C. elegans*. Neuron 16:35–46

Wang Q, Wadsworth WG (2002) The C domain of netrin UNC-6 silences calcium/calmodulin-dependent protein kinase- and diacylglycerol-dependent axon branching in *Caenorhabditis elegans*. J Neurosci 22:2274–2282

Watari-Goshima N, Ogura K, Wolf FW, Goshima Y, Garriga G (2007) *C. elegans* VAB-8 and UNC-73 regulate the SAX-3 receptor to direct cell and growth-cone migrations. Nat Neurosci 10:169–176

Wightman B, Clark SG, Taskar AM, Forrester WC, Maricq AV, Bargmann CI, Garriga G (1996) The C. elegans gene vab-8 guides posteriorly directed axon outgrowth and cell migration Development 122:671–682

Yonekawa Y, Harada A, Okada Y, Funakoshi T, Kanai Y, Takei Y, Terada S, Noda T, Hirokawa N (1998) Defect in synaptic vesicle precursor transport and neuronal cell death in KIF1A motor protein-deficient mice. J Cell Biol 141:431–441

Yu TW, Bargmann CI (2001) Dynamic regulation of axon guidance. Nat Neurosci Suppl 4:1169–1176

Yu TW, Hao JC, Lim W, Tessier-Lavigne M, Bargmann CI (2002) Shared receptors in axon guidance: SAX-3/Robo signals via UNC-34/Enabled and a Netrin-independent UNC-40/DCC function. Nat Neurosci 5(11):1147–1154

Yuan W, Zhou L, Chen JH, Wu JY, Rao Y, Ornitz DM (1999) The mouse SLIT family: secreted ligands for ROBO expressed in patterns that suggest a role in morphogenesis and axon guidance. Dev Biol 212:290–306

Zahn T, Angleson J, MacMorris M, Domke E, Hutton J, Schwartz C, Hutton J (2004) Dense core vesicle dynamics in *Caenorhabditis elegans* neurons and the role of kinesin UNC-104. Traffic 5:544–559

Zallen JA, Yi BA, Bargmann CI (1998) The conserved immunoglobulin superfamily member SAX-3/Robo directs multiple aspects of axon guidance in *C. elegans*. Cell 92(2):217–227

Chapter 4
Signaling Pathways Downstream of the Guidance Cues and Receptors

Abstract During nervous system development, the migration of cells and axons is a highly complex and orchestrated process, involving different steps. For the directional growth cone migration, signal is sensed from the extracellular cues, perceived by the receptors expressed in the growth cones, resulting in polarization of the growth cones followed by directional movement. Polarizing growth cones involves asymmetric localization of receptors on the growth cone membrane initiated by initial receptor and guidance cue interaction. In the second step, the growth cones extend in the direction set by the asymmetric localization of receptors. During the extension process, growth cones might encounter new cues, negatively or positively regulating the growth cone turning, branching, and synaptogenesis. Research in the past has supported the role of cytoskeleton polymers during cell migration and axon outgrowth, asymmetric accumulation of F-actin in the growth cone determining the direction of growth cone migration (Gallo et al. 2004; Lin and Forscher 1993; O'Connor and Bentley 1993). This chapter discusses the signaling molecules downstream of guidance receptor, linking the signal perceived at the cell surface to the actin modulatory proteins. Various molecules have been implicated working downstream of receptors including adaptor proteins, serving as platforms for protein–protein interactions, Rho GTPases and their effector molecules, and cytoskeleton modulatory protein.

Keywords Axon outgrowth · cell migration · RhoGTPases · AGE-1 · Rac effectors · Actin binding proteins

4.1 Molecules Acting Downstream of Receptors

The key signaling molecules identified downstream of guidance receptors include the Rac GTPase (belong to Rho GTPase family), calcium, and phosphoinositide 3-kinase (PI3K) pathways (Chang et al. 2006; Lundquist 2003; Ming et al. 1999). Cell migration studies in neutrophils have specified that interactions between these pathways are

A. Pandey and G. K. Pandey, *The UNC-53-mediated Interactome*,
SpringerBriefs in Neuroscience, DOI: 10.1007/978-3-319-07827-4_4,
© The Author(s) 2014

important for the establishment of polarity in response to chemotactic signal (Weiner 2002). Although the Rac and phosphoinositide signaling pathways have been implicated downstream of multiple guidance receptors, this section focuses on events downstream of the UNC-6/Netrin and its receptor UNC-40/DCC/FRA.

4.1.1 Rho Family GTPases

Rho family GTPases, belong to Ras superfamily of GTPases, regulating actin cytoskeleton from yeast to humans, acting as "molecular switches" by cycling between an inactive GDP bound state and an active GTP bound state, once activated they interact with variety of effector molecules to affect various biological processes (Hall 1998). In vertebrates, twenty Rho GTPase family members have been identified. The major role of Rho GTPases, conserved in all eukaryotes, is to control the cytoskeletal dynamics, cell movement, and other cellular processes. Three members of the family have been extensively studied, including Cdc42 (Cell division cycle42) known to affect filopodia dynamics, Rac1 (Ras-related C3 botulinum toxin substrate) affecting lamellipodia dynamics, and RhoA (Ras homologue gene family, member A) affects stress fibers. It is not surprising, therefore, that Rho GTPases play crucial roles in cell migration, axon outgrowth, guidance, branching, and fasciculation (Nobes and Hall 1995; Luo et al. 1996). Each Rho protein transduces the signal by interacting with numerous downstream target molecules collectively called the effectors, to bring about cytoskeleton reorganization in the GC.

In mammals, Rac1 and Cdc42 function during embryonic commissure axon outgrowth in response to Netrin-1 was evidenced by in vitro studies using N1E-115 neuroblastoma cell lines, where both Rac1 and Cdc42 activities were required for DCC-induced neurite outgrowth in response to Netrin (Lamoureux et al. 1997). Furthermore, this was achieved by Rac1-dependent actin reorganization (Li et al. 2002a, b). A correlation was established between DCC and Rac1 function-dependent filopodia formation, placing Rac downstream of DCC in NG108-15 cells, where a dominant-negative (D-N) Rac1 (T17 N) blocked the increase in DCC-dependent filopodia formation (Bourne et al. 1991; Shekarabi and Kennedy 2002). RhoA function was assessed, by transfecting constitutively active (CA) (G12 V) and D-N transgene in PC12 cell lines, the CA allele promoted and D-N allele inhibited neurite initiation and branching (Bourne et al. 1991; Sebök et al. 1999). Similarly, *Drosophila* genome encodes three rac genes: *RAC1, RAC2,* and *MTL* (Mig-two-like), sharing overlapping functions to control epithelial morphogenesis, myoblast fusion, and axon outgrowth and guidance (Hakeda-Suzuki 2002), both CA and D-N Rac alleles induced the disruption of axon outgrowth (Luo et al. 1994).

Caenorhabditis elegans Rho family GTPases include one Rho (*rho-1*), one Cdc42 (*cdc-42*), two canonical Rac1-like genes (*ced-10, rac-2*), and one Mtl Rac (*mig-2*) (Lundquist et al. 2001). Studies have shown that RhoG might be equivalent to Mtl/ MIG-2 in vertebrates (deBakka et al. 2004). Genetic analysis suggested that *ced-10* and *mig-2* act redundantly to control several developmental processes, including CAN axon guidance, phagocytosis of apoptotic cells, migration of DTCs and P cells, and axon outgrowth in D-type motoneurons (Zipkin et al. 1997; Lundquist et al. 2001; Spencer et al. 2001; Wu et al. 2002). A role for Rho GTPases in GC migration in worms was shown by expressing Rac and Cdc-42 CA transgene, G12 V, in PDE neuron, promoting ectopic lamellipodial and filopodial protrusions in the PDE neurons (Knobles et al. 1999; Alan et al. 2013). Phenotypic analysis of worms with MYR::UNC-40, a CA transgene in *ced-10* and *rac-2* background, revealed that Ced-10 and Rac-2 functions downstream of UNC-40/DCC in response to UNC-6/Netrin as they suppressed the MYR::UNC-40 phenotype. Biochemical analysis supported physical interaction between CED-10 and UNC-40 through the P2 cytoplasmic motif (Gitai et al. 2003).

4.1.2 Regulators of Rho GTPases

The activity of small GTPases depends on the ratio of the cytosolic GTP/GDP bound forms, determined by three classes of regulatory proteins including guanine nucleotide exchange factors (GEFs) which promote Rho GTPases to an active conformation, GTPase-activating proteins (GAPs) facilitate inactivation of GTPases by increasing the intrinsic GTPase activity, and guanine nucleotide dissociation inhibitors (GDIs), which bind to the GDP form of GTPases and prevent GDP release promoting the inactive state of the protein (D'Souza-Schorey et al. 1997; Kaibuchi et al. 1999).

GEFs catalyze the exchange of GDP to GTP on the Rho proteins and are composed of Dbl (diffusible B cell lymphoma) homology (DH) domain and pleckstrin homology (PH) domain, where the former is responsible for exchange activity and later for targeting function (Luo 2000; Montell 1999; Zheng 2001).

Numerous lines of evidence implicate various GEF molecules with neurite induction in vertebrates and flies. TIAM1 (the invasion inducing T-lymphoma and metastasis 1), STEF (SIF and TIAM 1 like exchange factor), FIR (FERM domain including RhoGEF), and PIX (PAK-interacting exchange factor) are vertebrate GEF proteins promoting neurite formation (Habets et al. 1994; Matsuo et al. 2002; Kubo et al. 2002). TIAM 1 and STEF are both involved in neurite extension in N1E115 cells (Matsuo et al. 2002; Leeuwen et al. 1997), FIR induces partial neurite retraction in cortical neurons (Kubo et al. 2002), and PIX is specific toward Rac and Cdc42, implicated in neurite outgrowth (Manser et al. 1998). Some RhoGEFs are composed of two GEF domains including TRIO, promoting neurite outgrowth, the first domain activates RhoG and Rac1 and second domain is active toward RhoA. Trio function in axon pathfinding is evident from *Drosophila trio* mutants, with misguided and prematurely terminating axons (Awasaki et al. 2000; Newsome et al. 2000). Kalirin is another GEF with two GEF domains, first GEF domain is specific to RhoG and Rac1 and second to RhoA, regulating filopodial neurite formation (Penzes et al. 2001). Additonally there are non-canonical GEF molecules including DOCK180 (Dedicator of cytokinesis), characterized by a "Docker" or "CZH2" domain instead of DH domain, shown to activate Rac by complexing with ELMO (Engulfment and cell motility) in cell migrations and engulfment (Lu and Ravichandran 2006). In vertebrates, DOCK180 was identified based on its interaction with the adaptor protein CRK (CT regulator of kinase), expression of DOCK180 in 3T3 fibroblasts caused cell surface extension phenotype (Hasegawa et al. 1996). DOCK180 is composed of SH2 and SH3 domains, and implicated in integrin-mediated signaling and cell movement (Mayer et al. 1988). *Drosophila*, DOCK180 homologue, MBC (Myoblast city), functions during myoblast fusion and epithelial cell migrations, and both processes require cell surface extension through actin cytoskeleton remodeling (Rushton et al. 1995; Erickson et al. 1997). DOCK180 might act as an effector for $3'$ PIs during axon guidance and cell migration. DOCK180 is a GEF activator of Rac that binds to and functions downstream of PtdIns(3,4,5)P3 during cell migration (Cote et al. 2005; Cote et al. 2007), suggesting that DOCK180 mediates a connection between the Rac and phosphoinositide signaling pathways.

The *C. elegans* genome encodes for nineteen Dbl-homology (DH) GEF proteins. UNC-73, a homologue of Trio, mutations in *unc-73* results in phenotypes including severe defects in axon extension, guidance, and fasciculation (Luo 2000; Montell 1999) and acts as a GEF for the MIG-2/RhoG (Kubiseski et al. 2003). UNC-73/Trio activity is required in many neurons during development including HSN, Q neuroblast, SM, and CAN (Steven et al. 1998). Genetic and biochemical evidence revealed that UNC-73/Trio, CED-2/CrkII (Chicken tumor virus no. 10 (CT10) regulator of kinase), and CED-5/DOCK180 act upstream of CED-10 and MIG-2 during outgrowth of D-type motoneuron axons (Wu et al. 2002). Recently, *C. elegans* homologue of PIX, PIX-1, has been shown to act as regulator of CED-10/Rac, acting redundantly to UNC-73/Trio, CED-10/Rac, UNC-73/Trio, and MIG-2/RhoG pathways in Q neuroblast protrusions and migrations (Dyer et al. 2010). *C. elegans*, TIAM-1, has been implicated in signaling downstream of UNC-40/DCC and as an activator of Cdc42 in neural protrusions and axon guidance (Demarco et al. 2012).

C. elegans homologue of DOCK180, CED-5 (cell death abnormal), was isolated as a gene required for phagocytosis of cell corpses and DTC migration, by extending cell surface for cell engulfment (Wu and Horvitz 1998), and CED-5 was shown to physically interact with CED-2/CrkII and function upstream of CED-10 in DTC migration and corpse engulfment (Reddien and Horvitz 2000). Genetic studies indicated that CED-5 functions in axon guidance (Lundquist et al. 2001). Consistent with this observation, a recent study reported, DOCK180 functions in response to Netrin/UNC-6 (Li et al. 2008). In addition to binding to PtdIns(3,4,5)P3, DOCK180 also associates with DCC, which might be mediated indirectly through the interaction between DOCK180 and PtdIns(3,4,5)P3. Moreover, both DCC and PtdIns(3,4,5)P3 are found within lipid rafts (Arcaro et al. 2007; Golub and Caroni 2005; Gomez-Mouton et al. 2004). Alternatively, it is possible that DOCK180 binds to both DCC and PtdIns(3,4,5)P3. The relationships between PtdIns(3,4,5)P3, DOCK180 and DCC are an important area for future investigations.

While GEF proteins are activators of Rho GTPases, GAP proteins function as inactivators or inhibitors. In vertebrates, several RhoGAP molecules have been identified and implicated in nervous system development including RICS (RhoGAP involved in β-catenin-N-cadherin receptor signaling), found in brain, inactivates

Cdc42 and Rac1, homozygous mutant mice for RICS have longer hippocampal and cerebellar granule neurite processes (Nasu-Nishimura et al. 2006), Grit, is mainly found in neuronal cells and is active toward RhoA and Cdc42, promotes neurite elongation in NGF (nerve growth factor)-stimulated cells (Nakamura et al. 2002), and p190RhoGAP, active toward RhoA, overexpression of p190RhoGAP causes extensive neurite outgrowth (Brouns et al. 2001). *C. elegans* genome encodes for twenty Rho GAPs, a study has reported SYD-1, a non-conventional GAP, with PDZ, C2, and RhoGAP like domains in neurite outgrowth and guidance supported by phenotypic analysis of worms with SYD-1 transgene lacking the RhoGAP domain, which interferes with neurite outgrowth and guidance in motoneurons (Hallam et al. 2002). Another non-conventional GAP and inhibitor of Rac *ced-10*, with no sequence similarity to known inhibitors, SWAN-1 (seven-WD-repeat protein of the AN11 family), was identified in a yeast two-hybrid screen with UNC-115/abLIM, an actin modulatory protein and is associated with RacGTPase. In mammalian fibroblasts, SWAN-1 loss of function suppressed defects caused by hypomorphic *ced-10* allele and enhanced lamellipodia and filopodia formation by CA Rac(G12 V), supporting that it negatively regulates *ced-10* (Yang et al. 2006).

4.1.3 Calcium

Calcium has emerged as an important second messenger downstream of guidance cues and receptors. Studies in *Xenopus* spinal neurons suggested that guidance cues such as Netrin-1 can generate a calcium gradients in the GC (Hong et al. 2000), resulting in local elevation of calcium levels, and this localized change in calcium level in turn can trigger activation of Rac and Cdc42 as studied in fetal rat hippocampal pyramidal neurons (Gomez and Zheng 2006; Jin et al. 2005). Role of calcium transient has been shown in axon outgrowth and branching of dissociated developing cortical neurons using fluorescence calcium imaging in response to Netrin-1, direct induction of localized calcium transient with photolysis of caged calcium-induced rapid outgrowth of axonal processes (Hutchin and Kalil 2008). Further, studies implicated calcium-/calmodulin-dependent kinase (CaMKII) and mitogen-activated protein kinase (MAPK) as downstream targets in this process (Tang and Kalil 2005). Netrin-1 and DCC regulate the migration and guidance of gonadotropin-releasing hormone (GnRH) neuronal GCs in calcium-dependent manner. Using calcium channel blockers such as nifedipine and ryanodine, which block extracellular L-type voltage-gated calcium channels (VGCC) and intracellular calcium channels, respectively, blocked the Netrin-1-induced GC motility (Low et al. 2012). FGF and its receptor FGFR have been implicated in increase in calcium flux leading to promotion in axon elongation (Archer et al. 1999).

Besides, acting in the Netrin/UNC-6 pathway, calcium acts downstream of Wnts. Wnt5a and its receptor Ryk repel callosal axons to cross the midline. In vitro studies established a correlation between the calcium transient frequency and axon outgrowth, knocking down Ryk inhibited calcium signaling and reduced axon outgrowth (Hutchins et al. 2012).

4.1.4 AGE-1/PI3K

Besides Rac and calcium, another cytosolic molecule implicated as downstream signal transducer is phosphoinositide 3-kinase (PI3 K). PI3 K is an enzyme that phosphorylates phosphoinositides at the $3'$ hydroxyl position to produce PtdIns $(3,4,5)P3$ and PtdIns $(3,4)P2$. PI3 K has been implicated in various cellular functions including cell growth, proliferation, differentiation, motility, survival, and intracellular trafficking (Waite and Eickholt 2010; Cain and Ridley 2009).

Insight into the role of $3'$ PIs in cell migration came from studies of chemotaxis in neutrophils, *Dictyostelium*, and fibroblasts (Funamoto et al. 2002; Haugh et al. 2000; Meili et al. 1999; Servant et al. 2000; Wang et al. 2002). Green fluorescent protein (GFP) fusions of pleckstrin homology (PH) domain-containing proteins bind to the products of PI3 K including PtdIns(3,4,5)$P3$ and PtdIns(3,4)$P2$ and preferentially localize to the leading edge of migrating cells suggesting that PI3 K functions through its products in response to chemoattractants. Inhibition of PI3 K activity resulted in inhibition of pseudopodia stability and actin polymerization at the leading edge, resulting in poor chemotactic response; moreover, PI3 K activity requires Rho GTPase function (Weiner et al. 2002). Genetic, pharmacological, and biochemical experiments have placed PI3 K both upstream and downstream of Rac in neutrophil chemotaxis (Weiner 2002; Servant et al. 2000; Benard et al. 1999; Hawkins et al. 1995). Based on these observations, existence of a positive feedback loop between Rac and $3'$ PIs has been predicted. Evidence for existence for such a loop comes from the fact that polarity in neutrophils is stimulated by direct delivery of exogenous $3'$ PIs, inhibitors of both PI3 K and Rho GTPases, can block this effect (Weiner et al. 2002).

Studies done in cultured *Xenopus* neurons suggested that PI3 K is required for Netrin response (Ming et al. 1999). Consistent with this observation, genetic studies in *C. elegans* indicate that AGE-1, an ortholog of PI3 K, functions downstream of MIG-10 in ventral process outgrowth in AVM neuron in response to UNC-6 and UNC-40 (Chang et al. 2006). AGE-1/PI3 K also works downstream of RhoGTPase Cdc42. Expression of CA transgene, Cdc42(G12V), in PDE neurons induces ectopic lamellipodia and filopodia protrusions, a phenotype suppressed by AGE-1/PI3 K, signaling through mTORC1 (mammalian target of rapamycin complex 1) (Alan et al. 2013).

4.1.5 UNC-34/Ena/VASP

The UNC-34/Ena (Enabled)/Mena (mammalian Ena)/VASP (vasodilator-stimulated phosphoprotein)/EVL (Enah/Vasp-like) is a family of actin-regulating proteins involved in nucleating actin polymerization in vitro (Huttelmaier et al. 1999; Lambrechts et al. 2000). In vivo Ena/VASP proteins are important for a number of actin-based cellular processes including axon guidance, phagocytosis, adhesion, cell polarization, platelet shape change, and VASP protein enhance actin-based motility of intracellular pathogen *Listeria monocytogenes* (Bear et al. 1990; Gertler et al. 1990, 1995; Lanier et al. 1999; Laurent et al. 1999; Vasioukhin et al. 2000). This protein family shares three conserved domains: an N-terminal Ena/VASP homology I domain (EVH1), a central proline-rich domain (PRD), and a C-terminal Ena/VASP homology II (EVH2) (Reinhard et al. 2001; Kwiatkowski et al. 2003). The EVH1 domain has been implicated in localization of UNC-34, indicated by the experiments done in epithelial cells (Fleming et al. 2010). The central PRD binds several known SH3 domain-containing proteins as well as the actin exchange factor profilin (Gertler et al. 1990, 1995; Reinhard et al. 1995, 2001). Finally, the C-terminal EVH2 domain binds G-actin and F-actin and is required for multimerization (Bachmann et al. 1999). *Drosophila* homologue, Ena was identified as a genetic suppressor of lethal mutations in tyrosine kinase, Abl (Gertler et al. 1990). Mutations in *ena* caused several axon guidance defects in the *Drosophila* NS (Gertler et al. 1995). Mena, the mammalian homologue, when enriched at the leading edge of fibroblast, decreased motility, while its depletion caused increase in motility, supporting that it functions in repulsion of GCs (Bear et al. 2000). *C. elegans unc-34* gene was identified as a genetic suppressor of ectopically expressed UNC-5 in touch neuron guidance, acting in the UNC-6/UNC-5 repulsive signaling pathway (Colavita and Culotti 1998). Analogous to vertebrates and fly homologue, UNC-34::GFP fusion protein has been localized to the leading edge of migrating epidermal cells during embryogenesis (Sheffield et al. 2007). UNC-34 acts in SLT-1-/SAX-3-mediated repulsive guidance in AVM ventral axon outgrowth, and in vitro binding assays supported that EVH1 domain of UNC-34 interacts with the SAX-3 CC2/CM2 motif (Fig. 4.1) (Yu et al. 2002). Besides, SAX-3 and UNC-34 also act in midline guidance similar to *Drosophila*, Robo and Enabled (Bashaw et al. 2000). In nematodes, UNC-34 also functions in attractive guidance pathway downstream of UNC-40 receptor. *unc-34* along with *ced-10* and *unc-115* was identified as suppressors of defects in AVM and HSN neurons, in gain-of-function, *mec-7::MYR::UNC-40* and *unc-86::MYR::UNC-40* transgenic worms. Genetic studies supported the model where CED-10 and UNC-115 act together in a parallel and redundant pathway to UNC-34, biochemical evidence supported physical ineraction between CED-10/Rac and UNC-115/abLIM with P2 motif of UNC-40, and UNC-34 with P1 motif of UNC-40 (Fig. 4.1) (Gitai et al. 2003). Based on a combination of genetic and RNAi experiments, UNC-34 function was implicated in morphogenesis in absence of *wsp-1* and *wve-1*, *C. elegans* homologues of WASP and WAVE, respectively (Withee at el. 2004). Additionally, the role UNC-34 downstream of receptors was supported by studying AQR and PQR axon pathfinding, placing UNC-34 in a parallel pathway to CED-10, RAC-2, and MIG-2 to regulate axon outgrowth and cell migration, probably by controlling formation of lamellipodia and filopodia (Shakir et al. 2006). Besides, it was shown that UNC-34 and cytoplasmic adaptor protein MIG-10/Lpd act during ventral process guidance of AVM and PVM neurons and promote lamellipodia formation and that UNC-34 is basically required for filopodia formation though the filopodia are dispensable for ventral guidance in HSN neurons, the GC advances toward ventral UNC-6 source by lamellipodia (Quinn et al. 2006; Chang et al. 2006). Recently, experiments done in PQR neuron have shown that UNC-34 along with UNC-115 and Arp2/3 acts in parallel pathways to initiate filopodia in GCs (Norris et al. 2009).

4.2 Rac and PI3K Effector Molecules

Effectors are proteins that bind to and mediate the downstream effects of activated Rac and 3′ PIs. Asymmetric localization of 3′ PIs, calcium, and activated Rac can lead to asymmetric recruitment of effectors for these molecules and set the direction of growth cone extension of cells and axons.

4.2.1 UNC-115/abLIM

unc-115 gene encodes an actin-binding protein, affecting axon guidance and outgrowth, and UNC-115 is similar to human actin-binding protein abLIM (actin-binding LIM protein1)/limatin, a tumor suppressor gene, composed of villin headpiece domain (VHD) and three LIM (Lin11, Isi-1, and Mec-3) domains that mediate protein–protein interactions (Struckhoff and Lundquist 2003; Roof et al. 1997). UNC-115 is expressed in neurons during development, and *unc-115* mutants are defective in dorsal guidance in all the DD and VD motor neurons, premature termination, and misrouting of sublateral nerves of SIADL, SMBDL, and SMDDR neurons (Lundquist et al. 1998). abLIM, the vertebrate homologue of UNC-115, has been implicated in retinal ganglion cell (RGC) axon pathfinding (Erkman et al. 2000; Lu et al. 2003); similarly, the fly homologue was also shown to be involved in axon projections in both visual and CNS (Garcia et al. 2007).

Genetic analysis in *C. elegans* confirmed that UNC-115 acts downstream of Racs. *unc-115* double mutants with *ced-10*, *mig-2*, or *unc-73* exhibited synthetic axon pathfinding defects in CAN and PDE neurons, and UNC-115 loss of function suppressed the formation of ectopic plasma membrane extensions induced by CA *RAC-2* (G12V) in neurons (Struckhoff and Lundquist 2003). Phenotypic analysis of gain-of-function UNC-40 molecule, *myr::UNC-40*, causing axon guidance defects including excess axon branching and excessive axon and cell body outgrowth, showed that reduction of UNC-115 function suppressed these defects acting through the P2 domain of UNC-40, this pathway also required CED-10 (Fig. 4.1) (Gitai et al. 2003). Experiments done in serum starved NIH 3T3 fibroblasts cells, exhibited that *myr::UNC-115* activity, membrane localization, and VHD promote lamellipodia and filopodia (Yang and Lundquist 2005). Recent studies in PDE neuron have shown that UNC-73 and UNC-115 act downstream of SAX-3 receptor in axon outgrowth, based on suppression of gain-of-function phenotype of *SAX-3::GFP* in PDE neurons (Li et al. 2013).

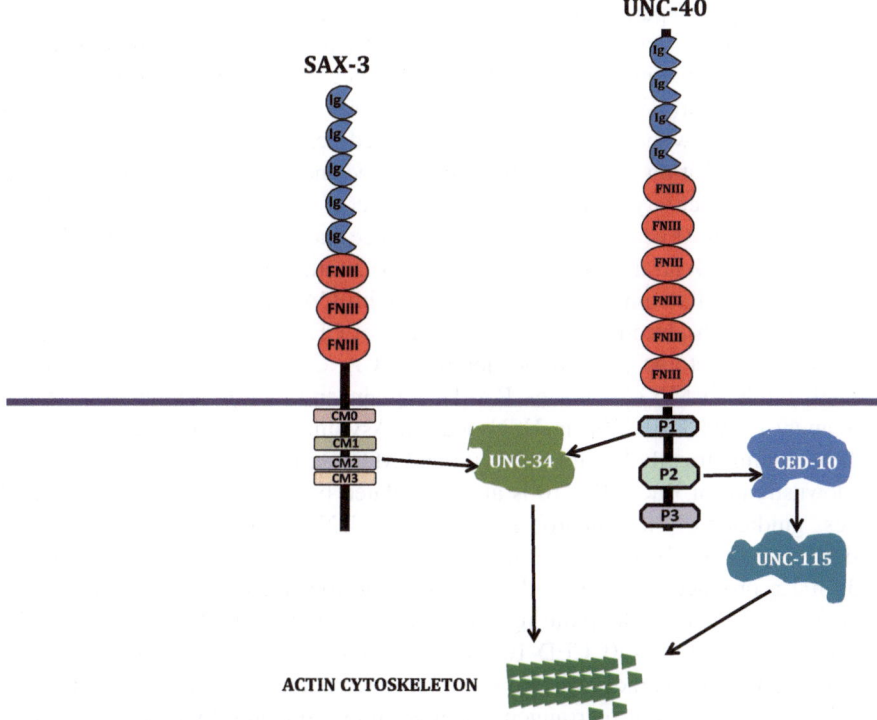

Fig. 4.1 Molecular mechanism of SLT-1/SAX-3, and UNC-6/UNC-40, signaling pathways: SLT-1/SAX-3, repulsive ligand receptor complex transduces signal through the actin-binding protein UNC-34 by direct physical interaction between the CM2/CC2 domain of SAX-3 and EVH1 domain of UNC-34. Similarly, UNC-6/UNC-40 attractive ligand receptor complex is linked to actin modulatory protein via two pathways including UNC-34 and CED-10/Rac. UNC-34 binds to the P1 conserved motif of UNC-40 receptor, whereas CED-10 transduces signals by physically interacting with the P2 conserved motif. UNC-115 acts downstream of CED-10 pathway. Both UNC-34 and UNC-115 are actin-binding proteins and provide a link between the actin cytoskeleton modulation and the receptor at the membrane

4.2.2 MIG-10/Lpd/RIAM

mig-10 encodes an actin-binding proteins that contains, from N- to C-terminus, an RAPH (Ras association pleckstrin homology) domain and a proline-rich motif, and MIG-10 is homologous to *Drosophila* Pico and vertebrate RIAM (Rap1-GTP-interacting adaptor molecule), Lpd (lamellipodin), and Grb7 (Growth factor receptor-binding protein 7), Grb10, and Grb14 constituting the MRL (MIG-10, RIAM, and Lpd) family of cytoplasmic adaptor proteins (Krause et al. 2004; Lafuente et al. 2004; Lyulcheva et al. 2008). MIG-10 and Lpd contain an EVH1-binding site which promotes binding to Ena/VASP family of proteins to promote actin filament elongation and oppose capping, though the evidence in worms supports MIG-10 and

UNC-34 acting in separate pathways (Chang et al. 2006; Drees and Gertler 2008; Krause et al. 2004; Quinn et al. 2006). In *C. elegans, mig-10* was identified in a screen for genes affecting CAN, ALM, and HSN neuron migrations during embryogenesis (Manser and Wood 1990; Manser et al. 1997). RIAM, the vertebrate homologue of MIG-10, when overexpressed in cultured cells, causes cell spreading and lamellipodia formation and overexpression of Lpd increases the velocity of actin-based protrusive activity in fibroblasts (Lafuente et al. 2004; Krause et al. 2004). In nematodes, MIG-10 acts downstream of guidance molecules UNC-6/UNC-40 and SLT-1/SAX-3 in ventral process guidance of HSN and AVM neurons basically by promoting filopodia formation. Epistasis test with CA *CED-10* transgene in PDE process outgrowth placed MIG-10 downstream of CED-10, since *mig-10* mutation suppressed the ectopic process phenotype of CA *CED-10* transgene. Moreover, CED-10 binds to MIG-10 via its RAPH domain and causes asymmetric ventral enrichment of MIG-10::YFP in HSN neuron. Asymmetric localization of MIG-10 leads to asymmetrically localized F-actin and MT (Quinn et al. 2008). Besides acting downstream of Rac, MIG-10 is also stimulated by AGE-1/PI3 K in the ventral process guidance of HSN neuron in response to UNC-6 and its receptor UNC-40 (Alder et al. 2006; Chang et al. 2006). The interaction of MIG-10 with CED-10 and PI(3,4)P2 a product of PI3 K and its asymmetric localization reminds of the Rac PI3 K positive feedback loop during chemotaxis of neutrophils (Weiner et al. 2002). Probably here too UNC-40, CED-10, AGE-1, and MIG-10 form a complex converting a shallow UNC-6 gradient into a sharp localized outgrowth-promoting activity.

MIG-10 brings about rearrangement of actin cytoskeleton by direct interaction with SH3 domain of ABI-1 (Abelson interactor-1), a member of WRC (WAVE regulatory complex). Phenotypic analysis of *abi-1 mig-10* double mutants in ALM migration and EC process outgrowth placed ABI-1 and MIG-10 in the same pathway. Epistasis test placed ABI-1 and WVE-1/WAVE downstream of MIG-10 in axon outgrowth pathways in response to UNC-6 and SLT-1 (Xu and Quinn 2012).

4.2.3 WAVE and WASP

The WASP (Wiskott-Aldrich syndrome protein), N-WASP (Neural-Wiskott-Aldrich syndrome protein), and SCAR/WAVE (WASP-like) proteins are actin cytoskeleton modulators ubiquitously found in eukaryotes (Kurisu and Takenawa 2009). The WASP and WAVE family proteins possess a C-terminal homologous sequence, the VCA region, consisting of the verprolin homology [also known as WASP homology 2 (WH2)] domain, the cofilin homology (also known as central) domain, and the acidic region, through which they bind to and activate the Arp2/3 complex. Besides, the VCA region, the WASP subfamily proteins are characterized by the N-terminal WH1 (WASP homology 1; also known as EVH1) domain, functioning in protein–protein interaction. In contrast, WAVE subfamily proteins have a N-terminus WHD/SHD domain (WAVE homology domain/SCAR homology domain) (Kurisu and Takenawa 2009).

Vertebrate genome encodes for five WASP family genes including *WASP*, *N-WASP*, *WAVE₁/SCAR₁*, *WAVE₂*, and *WAVE₃* (Stradal and Scita 2006; Takenawa and Suetsugu 2007). One member of each is found in *C. elegans*, *wve-1*, and *wsp-1* (Withee et al. 2004; Sawa et al. 2003). *Drosophila* has *SCAR* and *WASP* (Ben-Yaacov et al. 2001; Zallen et al. 2002). Both WASP and WAVE can associate directly or indirectly with membrane phosphoinositides.

N-WASP and WAVE were both identified as Rac effector's promoting filopodia and lamellipodia, respectively, involved in neuronal cell migration (Miki et al. 1997; Nozumi et al. 2003). When activated by Rac, the WAVE complex interacts with monomeric actin and the ARP2/3 complex to promote actin nucleation (Eden et al. 2002; Innocenti et al. 2004; Shakir et al. 2008). WAVE/SCAR activity is regulated by a plasma membrane-associated complex, and WRC (WAVE regulating complex) composed of Sra-1/PIR121, HEM2/NAP1/Kette, Abi1, and HSPC300 (Eden et al. 2002; Innocenti et al. 2004; Steffen et al. 2004; Stradal et al. 2004; Vartianien and Machesky 2004). In *C. elegans, gex-2* encodes Sra-1-like molecule and *gex-3* encodes kette-like molecule (Soto et al. 2002). Loss of WAVE function causes axonal defects in *Drosophila*, including ectopic midline crossing in the CNS and ectopic branching of motor neurons (Schenck et al. 2004; Zallen et al. 2002). The *C. elegans* WAVE homologue WVE-1 was identified in a screen for synthetic lethal mutations in *unc-34* null mutant background. Both WVE-1 and WSP-1 were shown to be required for CAN neuron migration and DD neuron dorsal process guidance; however, role for WSP-1 in neuronal migration was evident only upon *wsp-1* RNAi in *unc-34* mutants (Withee et al. 2004).

In *C. elegans*, evidence supports WASP and WAVE acting as Rac effectors redundantly in PDE axon guidance, where WAVE activity is controlled by CED-10. WVE-1/WAVE and CED-10/Rac act in parallel to a pathway containing WSP-1/WASP and MIG-2/RhoG, to activate Arp2/3 complex. GEX-2/Sra-1 and GEX-3/Kette, molecules that control WAVE activity, might act in both pathways. Furthermore, results showed that the CED-10/WVE-1 and MIG-2/WSP-1 pathways act in parallel to two other molecules known to control lamellipodia and filopodia formation, UNC-115/abLIM and UNC-34/Enabled (Shakir et al. 2008).

4.2.4 PAKs

PAKs are a family of serine threonine protein kinases related to Ste20 kinases, involved in diverse cellular functions including cell motility, cytoskeletal rearrangement, and filopodia formation. They were initially discovered as effectors of Rac and Cdc42 in a screen for specific Rho GTPase-binding partners (Manser et al. 1994). PAKs contain an N-terminal regulatory domain and a C-terminal catalytic domain. The regulatory domain is composed of a canonical (PXXP) and a non-canonical (PXP) SH3-binding motifs, CRIB (Cdc-42 and Rac interactive binding domain), PBD (p21 binding domain), and a $G_{\beta\gamma}$ subunit of heterotrimeric G protein-binding domain (Leeuw et al. 1998; Wang et al. 1999). Mammalian

Paks are divided into two groups: Group A (Pak1, 2, and 3) and Group B (Pak4, 5, and 6), whereas *C. elegans* has three PAK proteins including PAK-1, PAK-2, and MAX-2 (Chen et al. 1996; Hofmann et al. 2004).

The p21-activated kinase (PAK) proteins are effectors for Rac that have been implicated in guidance (Fan et al. 2003; Hing et al. 1999; Lucanic et al. 2006), transducing the signal by phosphorylating LIM kinase and Myosin Light Chain Kinase (MLCK), both regulators of actin dynamics (Linseman and Loucks 2008). LIM kinases are serine kinases implicated in the regulation of actin cytoskeletal dynamics through their ability to specifically phosphorylate members of the cofilin/actin depolymerizing factor (ADF) family. Studies have also placed PAKs upstream of Rac, and Pak2 phosphorylates and stimulates two GEF molecules βPIX and smgGDF to stimulate activation of Rac (Shin et al. 2006).

MAX-2 and PAK-1 in *C. elegans* have been implicated in DTC migration in Rac-dependent and Rac-independent pathways, respectively, and MAX-2 requires CED-10 function whereas PAK-1 requires PIX-1 (PAK-interacting exchange factor, a GEF) and GIT-1 (G protein-coupled receptor kinase interactor), a scaffolding protein to regulate DTC migration (Lucanic and Cheng 2008; Peters et al. 2013). MAX-2 and PAK-1 have been also implicated in P cell migration and guidance of VCCMNs (ventral cord commissural motoneurons) axons in response to UNC-6 and its UNC-5 receptor, acting downstream of Rac GTPases, CED-10 and MIG-2. Genetic evidence confirmed MAX-2 to function in pathway independent of Rac (Lucanic et al. 2006).

4.2.5 RACK-1

Rack1, receptor for activated C kinase1, was identified as a protein kinase C (PKC) interactor, mediating its plasma membrane translocation (Ron et al. 1994; Besson et al. 2002), composed of seven WD repeats, forming a seven-bladed beta propeller structure serving as a scaffold for protein–protein interactions (Neer et al. 1994). Both human and fly genome encode for single RACK1 proteins (Kadrmas et al. 2007). Studies showed that Rack1 regulates cell motility through its interaction with the Src tyrosine kinase, latter are known for their regulatory role in cell migration and focal adhesion (Besson et al. 2002; Lilienta and Chang 1998). Rack1 acts as a substrate for and inhibitor of Src kinase in response to active PKC (Besson et al. 2002; Schechtman and Mochly-Rosen 2001; Chang et al. 1998, 2002; Buensucesco et al. 2001). Expression of GFP tagged Rack with C-terminus truncated and a point mutation (Y246F), abolished Src binding in CHO-K1 cells, and inhibited cell protrusions and chemotactic migration (Cox et al. 2003). Moreover, Rack1 inhibited Src-induced cell motility in cultured 3T3 fibroblasts and Src mediated phosphorylation of p190RhoGAP (Miller et al. 2004), a modulator of Rho GTPase signaling and actin organization.

C. elegans, RACK-1 has been shown to be involved in embryonic cytokinesis, regulating recycling endosome distribution and membrane trafficking, and polarity stabilization in embryos (Ai et al. 2009, 2011). Using yeast two-hybrid, *C. elegans* RACK-1 was identified as a physical interactor of UNC-115/abLIM, binding between the LIM (N-terminus) and VHD (C-terminus) domains in the central region of UNC-115, further supported by co-immunoprecipitation assays (Lundquist et al. 1998). UNC-115/abLIM, an actin-binding protein, controls lamellipodial and filopodial formation in *C. elegans* GCs (Norris et al. 2009). RACK-1 loss-of-function data suggest it is necessary for axon guidance including VD/DD GABAergic motoneuron commissural axon guidance and DTC migration (Demarco and Lundquist 2011). A genetic epistasis analysis with CA form of *ced-10* (G12 V) (Struckhoff and Lundquist 2003) and *myr::UNC-115* (Yang and Lundquist 2005) was done to study the effect of loss of *rack-1* function. Loss of *rack-1* partially suppressed the gain-of-function defects of *ced-10* indicating that RACK-1 functions downstream of CED-10 but did not suppress the defects caused by activated UNC-115, placing RACK-1 upstream of UNC-115/abLIM (Demarco and Lundquist 2011). Both the activated forms of CED-10/Rac and UNC-115/abLIM caused formation of ectopic filopodia and lamellipodia protrusive structures in vivo in neurons (Demarco and Lundquist 2011; Yang and Lundquist 2005). Double mutant analysis placed RACK-1 in a pathway parallel to MIG-2/RhoG in axon pathfinding (Yang and Lundquist 2003; Struckhoff and Lundquist 2003).

4.2.6 MIG-15/NIK

The *mig-15* encodes the *C. elegans* homologue of the vertebrate Nck (non-catalytic region of tyrosine kinase adaptor protein 1)-interacting kinase (NIK), and the *Drosophila* Misshapen (Msn) protein, was identified as the gene required for DD and VD axon guidance (Poinat et al. 2002). Misshapen of *Drosophila* regulates cell migration and axon pathfinding, by signaling via a JNK/MAPK kinase pathway (Houalla et al. 2005; Su et al. 1998, 2000). Vertebrate studies have also implicated NIK kinases in cell migration, acting with JNK signaling (Xue et al. 2001). MIG-15 contains an N-terminal Pak/Ste20-like serine/threonine kinase domain followed by a proline-rich region, and a regulatory C-terminal citron-NIK homology (CNH) domain, which acts as Rho GTPase effector domain (Madaule et al. 2000). C-terminal region of MIG-15 interacts with the cytoplasmic domain of the β-integrin subunit INA-1 and Rac GTPases, and genetic studies place the two molecules in the same pathway in VD/DD GABAergic motor neuron axon pathfinding and Q descendent migration (Poinat et al. 2002). RNA-mediated interference (RNAi) of *ced-10*, *rac-2*, and *mig-2* enhanced a weak *mig-15* mutation (Poinat et al. 2002), indicating that MIG-15 acts downstream of all the three *C. elegans* Racs in axon pathfinding of the VD/DD, HSN, AVM, PVM, and PDE axon guidance (Shakir et al. 2006; Teulière et al. 2011). Role for MIG-15 downstream of Racs was supported by a genetic suppressor screen for ectopic filopodia and lamellipodia protrusions phenotype of CA *cdc-42* (G12V) transgene in PDE neurons, where *mig-15* loss of function suppressed the phenotype (Alan et al. 2013). Analogous to Misshapen, the fly homologue, which acts downstrem of Wnt pathway molecules including Frizzled and Disheveled (Paricio et al. 1999), MIG-15 was also shown to act upstream of MAB-5 in Wnt signaling pathway regulating posterior QL neuroblast direction of migration, evident from the phenotypic analysis of *mig-15 mab-5* double mutant, which was same as *mab-5* alone (Shakir et al. 2006).

Further studies have shown that MIG-15 is required for Q cell polarization and migration (Chapman et al. 2008). The role played by MIG-15 in cell migration and axon guidance is probably due to actin cytoskeleton remodeling. In *C. elegans,* a genetic enhancer screen using RNAi approach in *mig-15* mutant background, identified the actin modulators ERM-1 (a *C. elegans* homologue of ERM) as downstream effector of MIG-15 in DD commissural axon guidance. Time-lapse microscopy supported the role for MIG-15 and ERM-1 activity at the rear edge (Teulière et al. 2011). Previously, treatment of cultured cells with EGF (epidermal growth factor) and PDGF (platelet-derived growth factor) led NIK kinase-dependent phosphorylation of ERM (ezrin/radixin/moesin) proteins responsible for actin organization and lamellipodia formation (Baumgartner et al. 2006). Moreover, MIG-15 could possibly act downstream of UNC-40 and UNC-5 guidance receptors in response to UNC-6/Netrin in both ventral and dorsal axon guidance, a role supported by MIG-15-dependent polarized localization of MIG-10. Previously, UNC-40 had been implicated in Rac activation and MIG-10 localization at the leading edge (Adler et al. 2006; Quinn et al. 2006, 2008).

4.2.7 NCK-1/NCK

Mammalian NCK adaptor protein, encoded by two genes *NCK1* or *NCKα* and *NCK2* or *NCKβ* (Chen et al. 1998; Li et al. 2001), is adaptor proteins mainly composed of Src homology domains: SH2 and SH3 (Pawson and Nash 2003). NCK1 is actin cytoskeleton regulator and functions downstream of Robo, DCC, and the Eph RTKs (receptor tyrosine kinases) (Buday et al. 2002; Li et al. 2002a, b; Fan et al. 2003; Holland et al. 1997). The *Drosophila* homologue, DOCK (Dreadlock), regulates photoreceptor R cell axon guidance and targeting and work with Trio/UNC-73 to activate downstream effector Pak (Newsome et al. 2000). The *C. elegans* homologue, NCK-1, plays role in PLM, AVM, and PVM process guidance and AVM cell migration (Mohamed and Chin-Sang 2011). It also functions downstream of VAB-1/Eph RTK to modulate actin cytoskeleton via WSP-1/WASP in cell migration and axon guidance (Mohamed et al. 2012).

4.2.8 CED-2/CRKII

Crk (proto-oncogene c-Crk or p38), a family of adaptor proteins characterized by presence of SH2 and SH3 domains for protein–protein interaction, mammalian Crk protein family consists of CrkI, CrkII, and CrkL (Antoku and Mayer 2009), *Drosophila* has a single homologue, dCRK. Crk adaptor proteins have been

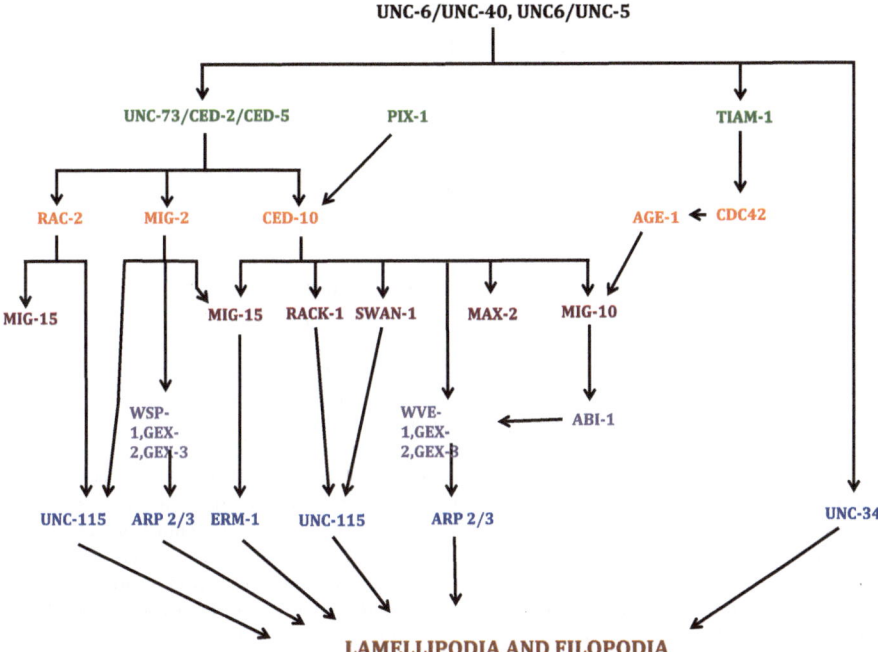

Fig. 4.2 Schematic of genetic pathways operating during cell migration and axon outgrowth in *C. elegans*: *Green text* represents the molecules which act as Rac regulators including the GEFs (Guanine nucleotide exchange factors), GAPs (GTPase-activating proteins), and GDI (Guanine nucleotide dissociation inhibitors); *red text* are Rac and Age-1 molecules which transduce signal downstream of receptors; *dark red text* represents effector molecules or molecules working downstream of Racs; *light purple text* represents ARP2/3 regulators; and *blue text* is for actin modulators or actin-binding proteins. In *C. elegans,* three Rac moecules: CED-10, RAC-2, and MIG-2 and one PI3 K, AGE-1 transduce signal from the receptors to their effector molecules (MIG-15, RACK-1, MIG-10, MAX-2, SWAN-1, and MIG-15), which transduce the signal either to ARP2/3 regulatory proteins including, WSP-1, WVE-1, and the WRC complex, composed of ABI-1, GEX-2, and GEX-3 or to the actin modulatory protein, which nucleate actin either *de novo* or from pre-existing actin filaments including UNC-115, ARP2/3, UNC-34, and ERM-1. All these pathways converge at the actin cytoskeleton reorganization to bring about growth cone movement

implicated in regulation of cellular adhesion, migration, and proliferation (Feller 2001). *C. elegans* homologue, CED-2/CrkII, is involved in cytoskeleton regulation, phagocytosis, and cell migration, by activating Racs (Antoku and Mayer 2009; Miyamoto and Yamauchi 2010). Studies have shown that CED-2/CrkII, CED-5/DOCK180, and CED-12/ELMO act genetically upstream of *C. elegans* Rac CED-10, to regulate the phagocytosis of apoptotic cells and the migration of DTCs (Wu and Horvitz 1998; Reddien and Horvitz 2000; Gumienny et al. 2001; Wu et al. 2001; Zhou et al. 2001).

A schematic of the genetic pathways downstream of various guidance molecules is depicted in (Figs. 4.2 and 4.3). These pathways are far from complete, and research in coming years will lead to identification of novel molecules which

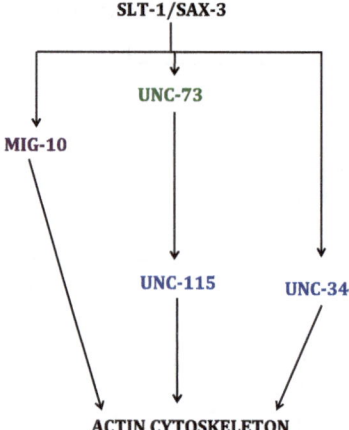

Fig. 4.3 Genetic pathways downstream of *C. elegans* SLT-1/SAX-3 during growth cone migration: *Green text* represents the molecules acting as Rac regulators including the GEFs (Guanine nucleotide exchange factors), GAPs (GTPase-activating proteins), GDI (Guanine nucleotide dissociation inhibitors); *red text* are Rac and Age-1 molecules which transduce signal downstream of receptors; *dark red text* represents effector molecules or molecules working downstream of Racs; *light purple text* represents ARP 2/3 regulators; and *blue text* is for actin modulators. Three pathways have been delineated transducing signals in response to SLT-1 comprised of MIG-10, UNC-73 (RacGEF) acting through UNC-115, and the actin-binding protein UNC-34

will fill up the gaps. The GCs decode the signal coming from the cue through the receptors and extend lamellipodia and filopodia, which powers the forward movement. Recently, a correlation has been shown in the extent of formation of protrusion in response to UNC-6 and UNC-40. Basically, UNC-6/Netrin controlled the polarity of protrusions and F-actin accumulation in response to attractive versus repulsive cues (Norris and Lundquist 2011).

References

Adler CE, Fetter RD, Bargmann CI (2006) UNC-6/Netrin induces neuronal asymmetry and defines the site of axon formation. Nat Neurosci 9:511–518

Ai E, Poole DS, Skop AR (2009) RACK-1 directs dynactin-dependent RAB-11endosomal recycling during mitosis in *Caenorhabditis elegans* Mol. Biol Cell 20:1629–1638

Ai E, Poole DS, Skop AR (2011) Long astral microtubules and RACK-1 stabilize polarity domains during maintenance phase in *Caenorhabditis elegans* embryos. PLoS One 6(4)

Alan JK, Struckhoff EC, Lundquist EA (2013) Multiple cytoskeletal pathways and PI3 K signaling mediate CDC-42-induced neuronal protrusion in *C. elegans*. Small GTPases 4(4):22

Antoku S, Mayer BJ (2009) Distinct roles for Crk adaptor isoforms in actin reorganization induced by extracellular signals. J Cell Sci 122(Pt 22):4228–4238

Arcaro A, Aubert M, Espinosa del Hierro ME, Khanzada UK, Angelidou S, Tetley TD, Bittermann AG, Frame MC, Seckl MJ (2007) Critical role for lipid raft-associated Src kinases in activation of PI3K-Akt signalling. Cell Signal. 19:1081–1092

Archer FR, Doherty P, Collins D, Bolsover SR (1999) CAMs and FGF cause a local submem-
brane calcium signal promoting axon outgrowth without a rise in bulk calcium concentra-
tion. Eur J Neurosci. 11(10):3565–3573

Awasaki T, Saito M, Sone M, Suzuki E, Sakai R, Ito K, Hama C (2000) The *Drosophila* trio
plays an essential role in patterning of axons by regulating their directional extension.
Neuron 26:119–131

Bachmann C, Fischer L, Walter U, Reinhard M (1999) The EVH2 domain of the vasodilator-
stimulated phosphoprotein mediates tetramerization, F-actin binding, and actin bundle formation. J Biol Chem 274(33):23549-23557

Bashaw GJ, Kidd T, Murray D, Pawson T, Goodman CS (2000) Repulsive axon guidance:
Abelson and Enabled play opposing roles downstream of the roundabout receptor. Cell
101(7):703–715

Baumgartner M, Sillman AL, Blackwood EM, Srivastava J, Madson N, Schilling JW, Wright JH,
Barber DL (2006) The Nck-interacting kinase phosphorylates ERM proteins for formation
of lamellipodium by growth factors. Proc Natl Acad Sci U S A. 103(36):13391–13396

Bear JE, Loureiro JJ, Libova I, Fassler R, Wehland J, Gertler FB (2000) Negative regulation of
fibroblast motility by Ena/ VASP proteins. Cell 101:717–728

Benard V, Bohl BP, Bokoch GM (1999) Characterization of *rac* and *cdc42* activation in chemoat-
tractant-stimulated human neutrophils using a novel assay for active GTPases. J Biol Chem
274:13198–13204

Ben-Yaacov S, Le Borgne R, Abramson I, Schweisguth F, Schejter ED (2001) Wasp, the
Drosophila Wiskott-Aldrich syndrome gene homologue, is required for cell fate decisions
mediated by Notch signaling. J Cell Biol 152:1–13

Besson A, Wilson TL, Yong VW (2002) The anchoring protein RACK1 links protein kinase
C epsilon to integrin beta chains. Requirements for adhesion and motility. J Biol Chem
277:22073–22084

Bourne HR, Sanders DA, McCormick F (1991) The GTPase superfamily: conserved structure
and molecular mechanism. Nature 349:117–127

Brouns MR, Matheson SF, Settleman J (2001) p190 RhoGAP is the principal Src substrate in
brain and regulates axon outgrowth, guidance and fasciculation. Nat Cell Biol 3:361–367

Buday L, Wunderlich L, Tamas P (2002) The Nck family of adapter proteins: regulators of actin
cytoskeleton. Cell Signal 14:723–731

Buensuceso CS, Woodside D, Huff JL, Plopper GE, O'Toole TE (2001) The WD protein Rack1
mediates protein kinase C and integrin-dependent cell Migration. J Cell Sci 114:1691–1698

Cain RJ, Ridley AJ (2009) Phosphoinositide 3-kinases in cell migration. Biol Cell. 101(1):13–29

Chang BY, Conroy KB, Machleder EM, Cartwright CA (1998) RACK1, a receptor for activated
C kinase and a homolog of the beta subunit of G proteins, inhibits activity of src tyrosine
kinases and growth of NIH 3T3 cells. Mol Cell Biol 18:3245–3256

Chang BY, Harte RA, Cartwright CA (2002) RACK1: a novel substrate for the Src protein-tyros-
ine kinase. Oncogene 21:7619–7629

Chang C, Adler CE, Krause M, Clark SG, Gertler FB, Tessier-Lavigne M, Bargmann CI (2006)
MIG-10/lamellipodin and AGE-1/PI3 K promote axon guidance and outgrowth in response
to slit and netrin. Curr Biol 16:854–862

Chapman JO, Li H, Lundquist EA (2008) The MIG-15 NIK kinase acts cell-autonomously in
neuroblast polarization and migration in *C. elegans*. Dev Biol 324(2):245–257

Chen W, Chen S, Yap SF, Lim L (1996) The *Caenorhabditis elegans* p21-activated kinase
(CePAK) colocalizes with CeRac1 and CDC42Ce at hypodermal cell boundaries during
embryo elongation. J Biol Chem 271:26362–26368

Chen HJ, Rojas-Soto M, Oguni A, Kennedy MB (1998) A synaptic Ras-GTPase activating protein (p135 SynGAP) inhibited by CaM kinase II. Neuron 20:895–904

Colavita A, Culotti JG (1998) Suppressors of ectopic UNC-5 growth cone steering identify eight
genes involved in axon guidance in *Caenorhabditis elegans*. Dev Biol 194:72–85

Côté JF, Vuori K (2007) GEF what? Dock180 and related proteins help Rac to polarize cells in
new ways. Trends Cell Biol 17:383–393

Côté JF, Motoyama AB, Bush JA, Vuori K (2005) A novel and evolutionarily conserved PtdIns(3,4,5) P3-binding domain is necessary for DOCK180 signaling. Nat Cell Biol 7:797–807

Cox EA, Bennin D, Doan AT, O'Toole T, Huttenlocher A (2003) RACK1 regulates integrin-mediated adhesion, protrusion, and chemotactic cell migration via its Src-binding site. Mol Biol Cell 14(2):658–669

deBakker CD, Haney LB, Kinchen JM, Grimsley C, Lu M, Klingele D, Hsu PK, Chou BK, Cheng LC, Blangy A, Sondek J, Hengartner MO, Wu YC, Ravichandran KS (2004) Phagocytosis of apoptotic cells is regulated by a UNC-73/TRIO-MIG-2/RhoG signaling module and armadillo repeats of CED-12/ELMO. Curr Biol 14:2208–2216

Demarco RS, Lundquist EA (2011) RACK-1 acts with Rac GTPase signaling and UNC-115/abLIM in *Caenorhabditis elegans* axon pathfinding and cell migration. PLoS Genet 6:1001215

Demarco RS, Struckhoff EC, Lundquist EA (2012) The Rac GTP exchange factor TIAM-1 acts with CDC-42 and the guidance receptor UNC-40/DCC in neuronal protrusion and axon guidance. PLoS Genet 8(4):

Drees F, Gertler FB (2008) Ena/VASP: proteins at the tip of the nervous system Curr. Opin Neurobiol 18:53–59

D'Souza-Schorey C, Boshans RL, McDonough M, Stahl PD, Van Aelst L (1997) A role for POR1, a Rac1-interacting protein, in ARF6-mediated cytoskeletal rearrangements. EMBO J 16(17):5445–5454

Dyer JO, Demarco RS, Lundquist EA (2010) Distinct roles of Rac GTPases and the UNC-73/Trio and PIX-1 Rac GTP exchange factors in neuroblast protrusion and migration in *C. elegans*. Small GTPases 1(1):44–61

Eden S, Rohatgi R, Podtelejnikov AV, Mann M, Kirschner MW (2002) Mechanism of regulation of WAVE1-induced actin nucleation by Rac1 and Nck. Nature 418(6899):790–793

Erickson M, Galletta BJ, Abmayr SM (1997) *Drosophila* myoblast city encodes a conserved protein that is essential for myoblast fusion, dorsal closure, and cytoskeletal organization. J Cell Biol 138:589–603

Erkman L, Yates PA, McLaughlin T, McEvilly RJ, Whisenhunt T, O'Connell SM, Krones AI, Kirby MA, Rapaport DH, Bermingham JR, O'Leary DD, Rosenfeld MG (2000) A POU domain transcription factor-dependent program regulates axon pathfinding in the vertebrate visual system. Neuron 28(3):779–792

Fan X, Labrador JP, Hing H, Bashaw GJ (2003) Slit stimulation recruits Dock and Pak to the roundabout receptor and increases Rac activity to regulate axon repulsion at the CNS midline. Neuron 40:113–127

Feller SM (2001) Crk family adaptors-signalling complex formation and biological roles. Oncogene 20:6348–6371

Fleming T, Chien SC, Vanderzalm PJ, Dell M, Gavin MK, Forrester WC, Garriga G (2010) The role of *C. elegans* Ena/VASP homolog UNC-34 in neuronal polarity and motility. Dev Biol 344(1):94–106

Funamoto S, Meili R, Lee S, Parry L, Firtel RA (2002) Spatial and temporal regulation of 3-phosphoinositides by PI 3-kinase and PTEN mediates chemotaxis. Cell 109:611–623

Gallo G, Letourneau PC (2004) Regulation of growth cone actin filaments by guidance cues. JNeurobiol 58:92–102

Garcia MC, Abbasi M, Singh S, He Q (2007) Role of *Drosophila* gene *dunc-115* in nervous system. Invert Neurosci 7(2):119–128

Gertler FB, Doctor JS, Hoffmann FM (1990) Genetic suppression of mutations in the *Drosophila* abl proto-oncogene homolog. Science 248:857–860

Gertler FB, Comer AR, Juang JL, Ahern SM, Clark MJ, Liebl EC, Hoffmann FM (1995) *enabled*, a dosage-sensitive suppressor of mutation in the *Drosophila* Abl tyrosine kinase, encodes an Abl substrate with SH3 domain-binding properties. Genes Dev 9:521–533

Gitai Z, Yu TW, Lundquist EA, Tessier-Lavigne M, Bargmann CI (2003) The netrin receptor UNC-40/DCC stimulates axon attraction and outgrowth through enabled and, in parallel, Rac and UNC-115/AbLIM. Neuron 9, 37(1):53–65

Golub T, Caroni P (2005) PI(4,5)P2-dependent microdomain assemblies capture microtubules to promote and control leading edge motility. J Cell Biol 169:151–165

Gomez TM, Zheng JQ (2006) The molecular basis for calcium-dependent axon pathfinding. Nat Rev Neurosci 7:115–125

Gómez-Moutón C, Lacalle RA, Mira E, Jiménez-Baranda S, Barber DF, Carrera AC, Martínez-A C, Mañes S (2004) Dynamic redistribution of raft domains as an organizing platform for signaling during cell chemotaxis. J Cell Biol 164:759–768

Gumienny TL, Brugnera E, Tosello-Trampont AC, Kinchen JM, Haney LB, Nishiwaki K, Walk SF, Nemergut ME, Macara IG, Francis R, Schedl T, Qin Y, Van Aelst L, Hengartner MO, Ravichandran KS (2001) CED-12/ELMO, a novel member of the CrkII/Dock180/Rac pathway, is required for phagocytosis and cell migration. Cell 107(1):27–41

Habets GG, Scholtes EH, Zuydgeest D, van der Kammen RA, Stam JC, Berns A, Collard JG (1994) Identification of an invasion-inducing gene, Tiam-1, that encodes a protein with homology to GDP-GTP exchangers for Rho like proteins. Cell 77:537–549

Hakeda-Suzuki S, Ng J, Tzu J, Dietzl G, Sun Y, Harms M, Nardine T, Luo L, Dickson BJ (2002) Rac function and regulation during Drosophila development. Nature 416(6879):438–442

Hall A (1998) Rho GTPases and the actin cytoskeleton. Science 279:509–514

Hallem SJ, Goncharov A, McEwen J, Baran R, Jin Y (2002) SYD-1, a presynaptic protein with PDZ, C2 and rhoGAP-like domains, specifies axon identity in C. elegans. Nat Neurosci 5(11):1137–1146

Hasegawa H, Kiyokawa E, Tanaka S, Nagashima K, Gotoh N, Shibuya M, Kurata T, Matsuda M (1996) DOCK180, a major CRK-binding protein, alters cell morphology upon translocation to the cell membrane. Mol Cell Biol 16:1770–1776

Haugh JM, Codazzi F, Teruel M, Meyer T (2000) Spatial sensing in fibroblasts mediated by 3′ phosphoinositides. J Cell Biol 151:1269–1280

Hawkins PT, Eguinoa A, Qiu RG, Stokoe D, Cooke FT, Walters R, Wennström S, Claesson-Welsh L, Evans T, Symons M, Stephens L (1995) PDGF stimulates an increase in GTP-Rac via activation of phosphoinositide 3- kinase. Curr Biol 5:393–403

Hing H, Xiao J, Harden N, Lim L, Zipursky SL (1999) Pak functions downstream of Dock to regulate photoreceptor axon guidance in Drosophila. Cell 97:853–863

Hofmann C, Shepelev M, Chernoff J (2004) The genetics of Pak. J Cell Sci 117:4343–4354

Holland SJ, Gale NW, Gish GD, Roth RA, Songyang Z, Cantley LC, Henkemeyer M, Yancopoulos GD, Pawson T (1997) Juxtamembrane tyrosine residues couple the Eph family receptor EphB2/Nuk to specific SH2 domain proteins in neuronal cells. EMBO J 16:3877–3888

Hong K, Nishiyama M, Henley J, Tessier-Lavigne M, Poo M (2000) Calcium signaling in the guidance of nerve growth by netrin-1. Nature 403:93–98

Houalla T, Hien Vuong D, Ruan W, Suter B, Rao Y (2005) The Ste20-like kinase misshapen functions together with Bicaudal-D and dynein in driving nuclear migration in the developing drosophila eye. Mech Dev. 122(1):97–108

Hutchins BI, Kalil K (2008) Differential outgrowth of axons and their branches is regulated by localized calcium transients. J Neurosci 28(1):143–153

Hutchins BI, Li L, Kalil K (2012) Wnt-induced calcium signaling mediates axon growth and guidance in the developing corpus callosum. Sci Signal 5(206):10

Huttelmaier S, Harbeck B, Steffens O, Messerschmidt T, Illen-berger S, Jockusch BM (1999) Characterization of the actin binding properties of the vasodilator-stimulated phosphoprotein VASP. FEBS Lett 451:68–74

Innocenti M, Zucconi A, Disanza A, Frittoli E, Areces LB, Steffen A, Stradal TE, Di Fiore PP, Carlier MF, Scita G (2004) Abi1 is essential for the formation and activation of a WAVE2 signalling complex. Nat Cell Biol 6:319–327

Jin M, Guan CB, Jiang YA, Chen G, Zhao CT, Cui K, Song YQ, Wu CP, Poo MM, Yuan XB (2005) Ca^{2+}-dependent regulation of rho GTPases triggers turning of nerve growth cones. J Neuroscience 25:2338–2347

Kadrmas JL, Smith AA, Pronovost SM, Beckerle MC (2007) Characterization of RACK1 Function in *Drosophila* Development. Dev Dyn 236:2207–2215

Kaibuchi K, Kuroda S, Amano M (1999) Regulation of the cytoskeleton and cell adhesion by the Rho family GTPases in mammalian cells. Annu Rev Biochem 68:459–486

Kim AC, Peters LL, Knoll JH, Huffel CV, Ciciotte SL, Kleyn Troemel ER, Chou JH, Dwyer ND, Colbert HA, Bargmann PW, Chisti AH (1997) Limatin (LIMAB), an actin-binding protein, maps to the mouse chromosome 19 and human chromosome 10q25, a region frequently deleted in human cancers. Genomics 46:291–293

Knobel KM, Jorgensen EM, Bastiani MJ (1999) Growth cones stall and collapse during axon outgrowth in *Caenorhabditis elegans*. Development 126:4489–4498

Krause M, Leslie JD, Stewart M, Lafuente EM, Valderrama F, Jagannathan R, Strasser GA, Rubinson DA, Liu H, Way M, Yaffe MB, Boussiotis VA, Gertler FB (2004a) Lamellipodin, an Ena/VASP ligand, is implicated in the regulation of lamellipodial dynamics. Dev Cell 7(4):571–583

Krause M, Leslie JD, Stewart M, Lafuente EM, Valderrama F, Jagannathan R, Strasser GA, Rubinson DA, Liu H, Way M, Yaffe MB, Boussiotis VA, Gertler FB (2004b) Lamellipodin, an Ena/VASP ligand, is implicated in the regulation of lamellipodial dynamics. Dev Cell 7:571–583

Kubiseski TJ, Culotti J, Pawson T (2003) Functional analysis of the *Caenorhabditis elegans* UNC-73B PH domain demonstrates a role in activation of the Rac GTPase in vitro and axon guidance in vivo. Mol Cell Biol 23(19):6823–6835

Kubo T, Yamashita T, Yamaguchi A, Sumimoto H, Hosokawa K, Tohyama M (2002) A novel FERM domain including guanine nucleotide exchange factor is involved in Rac signaling and regulates neurite remodeling. J Neurosci 22:8504–8513

Kurisu S, Takenawa T (2009) The WASP and WAVE family proteins. Genome Biol 10(6):226.1–226.9

Kwiatkowski AV, Gertler FB, Loureiro JJ (2003) Function and regulation of Ena/VASP proteins. Trends Cell Biol 13(7):386–392

Lafuente EM, van Puijenbroek AA, Krause M, Carman CV, Freeman GJ, Berezovskaya A, Constantine E, Springer TA, Gertler FB, Boussiotis VA (2004) RIAM, an Ena/VASP and Profilin ligand, interacts with Rap1-GTP and mediates Rap1-induced adhesion. Dev Cell 7:585–595

Lambrechts A, Kwiatkowski AV, Lanier LM, Bear JE, Vande-kerckhove J, Ampe C, Gertler FB (2000) cAMP-dependent protein kinase phosphorylation of EVL, a Mena/VASP relative, regu- lates its interaction with actin and SH3 domains. J Biol Chem 275:36143–36151

Lamoureux P, Altun-Gultekin ZF, Lin C, Wagner JA, Heidemann SR (1997) Rac is required for growth cone function but not neurite assembly. J Cell Sci 110(Pt 5):635–641

Lanier LM, Gates MA, Witke W, Menzies AS, Wehman AM, Macklis JD, Kwiatkowski D, Soriano P, Gertler FB (1999) Mena is required for neurulation and commissure formation. Neuron 22(2):313–325

Laurent V, Loisel TP, Harbeck B, Wehman A, Gröbe L, Jockusch BM, Wehland J, Gertler FB, Carlier MF (1999) Role of proteins of the Ena/VASP family in actin-based motility of Listeria monocytogenes. J Cell Biol. 144(6):1245–1258

Leeuw T, Wu C, Schrag JD, Whiteway M, Thomas DY, Leberer E (1998) Interaction of a G-protein beta-subunit with a conserved sequence in Ste20/PAK family protein kinases. Nature 8:391

Leeuwen FN, Kain HE, Kammen RA, Michiels F, Kranenburg OW, Collard JG (1997) The gua- nine nucleotide exchange factor Tiam1 affects neuronal morphology; opposing roles for the small GTPases Rac and Rho. J Cell Biol 139:797–807

Li W, Fan J, Woodley DT (2001) Nck/Dock: an adapter between cell surface receptors and the actin cytoskeleton. Oncogene 20:6403–6417

Li X, Meriane M, Triki I, Shekarabi M, Kennedy TE, Larose L, Lamarche-Vane N (2002a) The adaptor protein Nck-1 couples the *netrin-1* receptor DCC (deleted in colorectal cancer) to the activation of the small GTPase Rac1 through an atypical mechanism. J Biol Chem 277:37788–37797

Li X, Saint-Cyr-Proulx E, Aktories K, Lamarche-Vane N (2002b) Rac1 and Cdc42 but not RhoA or Rho kinase activities are required for neurite outgrowth induced by the Netrin-1 receptor DCC (deleted in colorectal cancer) in N1E-115 neuroblastoma cells. J Biol Chem 277:15207–15214

Li H, Kulkarni G, Wadsworth WG (2008a) RPM-1, a *Caenorhabditis elegans* protein that functions in presynaptic differentiation, negatively regulates axon outgrowth by controlling SAX-3/robo and UNC-5/UNC5 activity. J Neurosci 28:3595–3603

Li X, Gao X, Liu G, Xiong W, Wu J, Rao Y (2008b) Netrin signal transduction and the guanine nucleotide exchange factor DOCK180 in attractive signaling. Nat Neurosci 11:28–35

Li J, Pu P, Le W (2013) The SAX-3 receptor stimulates axon outgrowth and the signal sequence and transmembrane domain are critical for SAX-3 membrane localization in the PDE neuron of *C. elegans*. PLoS One 8(6):12

Liliental J, Chang DD (1998) Rack1, a receptor for activated protein kinase C, interacts with integrin beta subunit. J Biol Chem 273(4):2379–2383

Lin CH, Forscher P (1993) Cytoskeletal Remodeling during growth cone-target interactions. J Cell Biol 121:1369–1383

Linseman DA, Loucks FA (2008) Diverse roles of Rho family GTPases in neuronal development, survival, and death. Front Biosci 13:657–676

Low VF, Fiorini Z, Fisher L, Jasoni CL (2012) Netrin-1 stimulates developing GnRH neurons to extend neurites to the median eminence in a calcium- dependent manner. PLoS ONE 7(10)

Lu M, Ravichandran KS (2006) Dock180-ELMO cooperation in Rac activation. Methods Enzymol 406:388–402

Lu C, Huang X, Ma HF, Gooley JJ, Aparacio J, Roof DJ, Chen C, Chen DF, Li T (2003) Normal retinal development and retinofugal projections in mice lacking the retina-specific variant of actin-binding LIM domain protein. Neuroscience 120(1):121–131

Lucanic M, Cheng HJ (2008) A RAC/CDC-42-independent GIT/PIX/PAK signaling pathway mediates cell migration in *C. elegans*. PLoS Genet 4(11)

Lucanic M, Kiley M, Ashcroft N, L'etoile N, Cheng HJ (2006) The *Caenorhabditis elegans* P21-activated kinases are differentially required for UNC-6/netrin-mediated commissural motor axon guidance. Development 133:4549–4559

Lundquist EA (2003) Rac proteins and the control of axon development. Curr Opin Neurobiol 13:384–390

Lundquist EA, Herman RK, Shaw JE, Bargmann CI (1998) UNC-115, a conserved protein with predicted LIM and actin-binding domains, mediates axon guidance in *C. elegans*. Neuron 21:385–392

Lundquist EA, Reddien PW, Hartwieg E, Horvitz HR, Bargmann CI (2001) Three *C. elegans* Rac proteins and several alternative Rac regulators control axon guidance, cell migration and apoptotic cell phagocytosis. Development 128:4475–4488

Luo L (2000) Rho GTPases in neuronal morphogenesis. Nat Rev Neurosci 1:173–180

Luo L, Jan L, Jan YN (1996) Small GTPases in axon outgrowth. Perspect Dev Neurobiol 4(2–3):199–204

Luo L, Liao YJ, Jan LY, Jan YN (1994) Distinct morphogenetic functions of similar small GTPases: Drosophila Drac1 is involved in axonal outgrowth and myoblast fusion. Genes Dev 8(15):1787–802

Lyulcheva E, Taylor E, Michael M, Vehlow A, Tan S, Fletcher A, Krause M, Bennett D (2008) *Drosophila* pico and its mammalian ortholog lamellipodin activate serum response factor and promote cell proliferation. Dev Cell 15:680–690

Manser J, Wood WB (1990) Mutations affecting embryonic cell migrations in *Caenorhabditis elegans*. Dev Genet 11(1):49–64

Manser E, Leung T, Salihuddin H, Zhao Z-S, Lim L (1994) A brain serine/threonine protein kinase activated by Cdc42 and Rac1. Nature 367:40–46

Manser J, Roonprapunt C, Margolis B (1997) *C. elegans* cell migration gene *mig-10* shares similarities with a family of SH2 domain proteins and acts cell nonautonomously in excretory canal development. Dev Biol 184(1):150–164

Manser E, Loo TH, Koh CG, Zhao ZS, Chen XQ, Tan L, Tan I, Leung T, Lim L (1998) PAK kinases are directly coupled to the PIX family of nucleotide exchange factors. Mol Cell 1(2):183–192

Matsuo N, Hoshino M, Yoshizawa M, Nabeshima Y (2002) Characterization of STEF, a guanine nucleotide exchange factor for Rac1, required for neurite growth. J Biol Chem 277:2860–2868

Mayer BJ, Hamaguchi M, Hanafusa H (1988) A novel viral oncogene with structural similarity to phospholipase C. Nature 332:272–275

Meili R, Ellsworth C, Lee S, Reddy TB, Ma H, Firtel RA (1999) Chemoattractant-mediated transient activation and membrane localization of Akt/PKB is required for efficient chemotaxis to cAMP in *Dictyostelium*. EMBO J 18:2092–2105

Miki H, Nonoyama S, Zhu Q, Aruffo A, Ochs HD, Takenawa T (1997) Tyrosine kinase signaling regulates Wiskott-Aldrich syndrome protein function, which is essential for megakaryocyte differentiation. Cell Growth Differ. 8:195–202

Miller LD, Lee KC, Mochly-Rosen D, Cartwright CA (2004) RACK1 regulates Src-mediated Sam68 and p190RhoGAP signaling. Oncogene 23:5682–5686

Ming G, Song H, Berninger B, Inagaki N, Tessier-Lavigne M, Poo M (1999) Phospholipase C-gamma and phosphoinositide 3-kinase mediate cytoplasmic signaling in nerve growth cone guidance. Neuron 23:139–148

Miyamoto Y, Yamauchi J (2010) Cellular signaling of Dock family proteins in neural function. Cell Signal. 22(2):175--182

Mohamed AM, Chin-Sang ID (2011) The *C. elegans* nck-1 gene encodes two isoforms and is required for neuronal guidance. Dev Biol 354:55–66

Montell DJ (1999) The genetics of cell migration in *Drosophila melanogaster* and *Caenorhabditis elegans* development. Development 126:3035

Nakamura T, Komiya M, Sone K, Hirose E, Gotoh N, Morii H, Ohta Y, Mori N (2002) Grit, a GTPase activating protein for the Rho family, regulates neurite extension through association with the TrkA receptor and N-Shc and CrkL/Crk adapter molecules. Mol Cell Biol 22:8721–8734

Nasu-Nishimura Y, Hayashi T, Ohishi T, Okabe T, Ohwada S, Hasegawa Y, Senda T, Toyoshima C, Nakamura T, Akiyama T (2006) Role of the Rho GTPase-activating protein RICS in neurite outgrowth. Genes Cells 11:607–614

Neer EJ, Schmidt CJ, Nambudripad R, Smith TF (1994) The ancient regulatory-protein family of WD-repeat proteins. Nature 371:297–300

Newsome TP, Schmidt S, Dietzl G, Keleman K, Asling B, Debant A, Dickson BJ (2000) Trio combines with dock to regulate Pak activity during photoreceptor axon pathfinding in *Drosophila*. Cell 101:283–294

Nobes CD, Hall A (1995) Rho, Rac and Cdc42 GTPases regulate the assembly of multi-molecular focal complex associated with actin stress fibers, lamellipodia and filopodia. Cell 81:53–62

Norris AD, Dyer JO, Lundquist EA (2009) The Arp2/3 complex, UNC-115/abLIM, and UNC-34/Enabled regulate axon guidance and growth cone filopodia formation in *Caenorhabditis elegans*. Neural Dev 4:38

Norris AD, Lundquist EA (2011) UNC-6/netrin and its receptors UNC-5 and UNC-40/DCC modulate growth cone protrusion in vivo in C. elegans. Development 138(20):4433--4442

Nozumi M, Nakagawa H, Miki H, Takenawa T, Miyamoto S (2003) Differential localization of WAVE isoforms in filopodia and lamellipodia of the neuronal growth cone. J Cell Sci. 116(Pt 2):239–246

O'Connor TP, Bentley D (1993) Accumulation of actin in subsets of pioneer growth cone filopodia in response to neural and epithelial guidance cues *in situ*. J Cell Biol 123:935–948

Paricio N, Feiguin F, Boutros M, Eaton S, Mlodzik M (1999) The *Drosophila* STE20-like kinase misshapen is required downstream of the Frizzled receptor in planar polarity signaling. EMBO J 18:4669–4678

Pawson T, Nash P (2003) Assembly of cell regulatory systems through protein interaction domains. Science 300:445–452

Penzes P, Johnson RC, Sattler R, Zhang X, Huganir RL, Kambampati V, Mains RE, Eipper BA (2001) The neuronal Rho-GEF Kalirin-7 interacts with PDZ domain-containing proteins and regulates dendritic morphogenesis. Neuron 29:229–242

Peters EC, Gossett AJ, Goldstein B, Der CJ, Reiner DJ (2013) Redundant canonical and noncanonical Caenorhabditis elegans p21-activated kinase signaling governs distal tip cell migrations. G3 (Bethesda) 3(2):181–195

Poinat P, De Arcangelis A, Sookhareea S, Zhu X, Hedgecock EM, Labouesse M, Georges-Labouesse E (2002) A conserved interaction between beta1 integrin/PAT-3 and Nck-interacting kinase/MIG-15 that mediates commissural axon navigation in C. elegans. Curr Biol. 12(8):622–631

Quinn CC, Pfeil DS, Chen E, Stovall EL, Harden MV, Gavin MK, Forrester WC, Ryder EF, Soto MC, Wadsworth WG (2006) UNC-6/netrin and SLT-1/slit guidance cues orient axon outgrowth mediated by MIG-10/RIAM/lamellipodin. Curr Biol 16:845–853

Quinn CC, Pfeil DS, Wadsworth WG (2008) CED-10/Rac1 mediates axon guidance by regulating the asymmetric distribution of MIG-10/lamellipodin. Curr Biol 18:808–813

Reddien PW, Horvitz HR (2000) CED-2/CrkII and CED-10/Rac control phagocytosis and cell migration in *Caenorhabditis elegans*. Nat Cell Biol 2(3):131–136

Reinhard M, Jarchau T, Walter U (2001) Actin-based motility: stop and go with Ena/VASP proteins. Trends Biochem Sci. 26(4):243–249

Reinhard M, Jouvenal K, Tripier D, Walter U (1995) Identification, purification, and characterization of a zyxin-related protein that binds the focal adhesion and microfilament protein VASP (vasodilator-stimulated phosphoprotein). Proc Natl Acad Sci U S A. 92(17):7956--7960

Ron D, Chen CH, Caldwell J, Jamieson L, Orr E, Mochly-Rosen D (1994) Cloning of an intracellular receptor for protein kinase C: a homolog of the beta subunit of G proteins. Proc Natl Acad Sci USA 91:839–843

Roof DJ, Hayes A, Adamian M, Chisti AH, Li T (1997) Molecular characterization of abLIM, a novel actin-binding and double zinc finger protein. J Cell Biol 138:575–588

Rushton E, Drysdale R, Abmayr SM, Michelson AM, Bate M (1995) Mutations in a novel gene, myoblast city, provide evidence in support of the founder cell hypothesis for *Drosophila* muscle development. Development 121:1979–1988

Sawa M, Suetsugu S, Sugimoto A, Miki H, Yamamoto M, Takenawa T (2003) Essential role of the *C. elegans* Arp2/3 complex in cell migration during ventral enclosure. J Cell Sci 116:1505–1518

Schechtman D, Mochly-Rosen D (2001) Adaptor proteins in protein kinase C-mediated signal transduction. Oncogene 20(44):6339–6347

Schenck A, Qurashi A, Carrera P, Bardoni B, Diebold C, Schejter E, Mandel JL, Giangrande A (2004) WAVE/SCAR, a multifunctional complex coordinating different aspects of neuronal connectivity. Dev Biol 274:260–270

Sebök A, Nusser N, Debreceni B, Guo Z, Santos MF, Szeberenyi J, Tigyi G (1999) Different roles for RhoA during neurite initiation, elongation, and regeneration in PC12 cells. J Neurochem 73(3):949–960

Servant G, Weiner OD, Herzmark P, Balla T, Sedat JW, Bourne HR (2000) Polarization of chemoattractant receptor signaling during neutrophil chemotaxis. Science 287:1037–1040

Shakir MA, Gill JS, Lundquist EA (2006) Interactions of UNC-34 Enabled with Rac GTPases and the NIK kinase MIG-15 in Caenorhabditis elegans axon pathfinding and neuronal migration. Genetics 172(2):893--913

Shakir MA, Jiang K, Struckhoff EC, Demarco RS, Patel FB, Soto MC, Lundquist EA (2008) The Arp2/3 activators WAVE and WASP have distinct genetic interactions with Rac GTPases in *Caenorhabditis elegans* axon guidance. Genetics 179(4):1957–1971

Sheffield M, Loveless T, Hardin J, Pettitt J (2007) C. elegans Enabled exhibits novel interactions with N-WASP, Abl, and cell-cell junctions. Curr Biol 17(20):1791–1796

Shekarabi M, Kennedy TE (2002) The netrin-1 receptor DCC promotes filopodia formation and cell spreading by activating Cdc42 and Rac1. Mol Cell Neurosci 19:1–17

Shin EY, Lee CS, Cho TG, Kim YG, Song S, Juhnn YS, Park SC, Manser E, Kim EG (2006) BetaPak-interacting exchange factor-mediated Rac1 activation requires smgGDS guanine nucleotide exchange factor in basic fibroblast growth factor-induced neurite outgrowth. J Biol Chem 281:35954–35964

Soto MC, Qadota H, Kasuya K, Inoue M, Tsuboi D, Mello CC, Kaibuchi K (2002) The GEX-2 and GEX-3 proteins are required for tissue morphogenesis and cell migrations in *C. elegans*. Genes Dev 16:620–632

Spencer AG, Orita S, Malone CJ, Han M (2001) A RHO GTPase-mediated pathway is required during P cell migration in Caenorhabditis elegans. Proc Natl Acad Sci U S A. 98(23):13132–13137

Steffen A, Rottner K, Ehinger J, Innocenti M, Scita G, Wehland J, Stradal TE (2004) Sra-1 and Nap1 link Rac to actin assembly driving lamellipodia formation. EMBO J 23:749–759

Steven R, Kubiseski TJ, Zheng H, Kulkarni S, Mancillas J, Ruiz Morales A, Hogue CW, Pawson T, Culotti J (1998) UNC-73 activates the Rac GTPase and is required for cell and growth cone migrations in *C. elegans*. Cell 92:785–795

Stradal TE, Scita G (2006) Protein complexes regulating Arp2/3-mediated actin assembly. Curr Opin Cell Biol 18:4–10

Stradal TE, Rottner K, Disanza A, Confalonieri S, Innocenti M, Scita G (2004) Regulation of actin dynamics by WASP and WAVE family proteins. Trends Cell Biol 14:303–311

Struckhoff EC, Lundquist EA (2003) The actin-binding protein UNC-115 is an effector of Rac signaling during axon pathfinding in *C. elegans*. Development 130:693–704

Su M, Merz DC, Killeen MT, Zhou Y, Zheng H, Kramer JM, Hedgecock EM, Culotti JG (2000) Regulation of the UNC-5 netrin receptor initiates the first reorientation of migrating distal tip cells in Caenorhabditis elegans. Development 127(3):585–594

Su YC, Treisman JE, Skolnik EY (1998) The Drosophila Ste20-related kinase misshapen is required for embryonic dorsal closure and acts through a JNK MAPK module on an evolutionarily conserved signaling pathway. Genes Dev. 12(15):2371–2380

Takenawa T, Suetsugu S (2007) The WASP-WAVE protein network: connecting the membrane to the cytoskeleton. Nat Rev Mol Cell Biol 8:37–48

Tang F, Kalik K (2005) Netrin-1 induces axon branching in developing cortical neurons by frequency-dependent calcium signaling pathways. J Neurosci 25(28):6702–6715

Teulière J, Gally C, Garriga G, Labouesse M, Georges-Labouesse E (2011) MIG-15 and ERM-1 promote growth cone directional migration in parallel to UNC-116 and WVE-1. Development 138(20):4475–4485

Vartiainen MK, Machesky LM (2004) The WASP-Arp2/3 pathway: genetic insights. Curr Opin Cell Biol 16:174–181

Vasioukhin V, Bauer C, Yin M, Fuchs E (2000) Directed actin polymerization is the driving force for epithelial cell-cell adhesion. Cell 100(2):209–219

Waite K, Eickholt BJ (2010) The neurodevelopmental implications of PI3K signaling. Curr Top Microbiol Immunol. 346:245–265

Wang J, Frost JA, Cobb MH, Ross EM (1999) Reciprocal signaling between heterotrimeric G proteins and the p21-stimulated protein kinase. J Biol Chem 274(44):31641–31647

Wang F, Herzmark P, Weiner OD, Srinivasan S, Servant G, Bourne HR (2002) Lipid products of PI(3)Ks maintain persistent cell polarity and directed motility in neutrophils. Nat Cell Biol 4:513–518

Weiner OD (2002) Regulation of cell polarity during eukaryotic chemotaxis: the chemotactic compass. Curr Opin Cell Biol 14:196–202

Weiner OD, Neilsen PO, Prestwich GD, Kirschner MW, Cantley LC, Bourne HR (2002) A PtdInsP(3)- and Rho GTPase-mediated positive feedback loop regulates neutrophil polarity. Nat Cell Biol 4:509–513

Withee J, Galligan B, Hawkins N, Garriga G (2004) *Caenorhabditis elegans* WASP and Ena/VASP Proteins Play Compensatory Roles in Morphogenesis and Neuronal Cell Migration. Genetics 167:1165–1176

Wu YC, Horvitz HR (1998) *C. elegans* phagocytosis and cell-migration protein CED-5 is similar to human DOCK180. Nature 392(6675):501–504

Wu YC, Cheng TW, Lee MC, Weng NY (2002) Distinct rac activation pathways control *Caenorhabditis elegans* cell migration and axon outgrowth. Dev Biol 250(1):145–155

Wu YC, Tsai MC, Cheng LC, Chou CJ, Weng NY (2001) *C. Elegans* CED–12 Acts in the conserved crkII/DOCK180/Rac pathway to control cell migration and cell corpse engulfment. Dev Cell 1(4):491–502

Xu Y, Quinn CC (2012) MIG-10 functions with ABI-1 to mediate the UNC-6 and SLT-1 axon guidance signaling pathways. PLoS Genet 8(11):e1003054

Xue Y, Wang X, Li Z, Gotoh N, Chapman D, Skolnik EY (2001) Mesodermal patterning defect in mice lacking the Ste20 NCK interacting kinase (NIK). Development 128(9):1559---1572

Yang Y, Lundquist EA (2005) The actin-binding protein UNC-115/abLIM controls formation of lamellipodia and filopodia and neuronal morphogenesis in *Caenorhabditis elegans* Mol. Cell Biol 25(12):5158–5170

Yang Y, Lu J, Rovnak J, Quackenbush SL, Lundquist EA (2006) SWAN-1, a *Caenorhabditis elegans* WD repeat protein of the AN11 family, is a negative regulator of Rac GTPase function. Genetics 174(4):1917–1932

Yang Y, Lundquist EA (2005) The actin-binding protein UNC-115/abLIM controls formation of lamellipodia and filopodia and neuronal morphogenesis in Caenorhabditis elegans Mol. Cell Biol. 25(12):5158–5170

Yu TW, Hao JC, Lim W, Tessier-Lavigne M, Bargmann CI (2002) Shared receptors in axon guidance: SAX-3/Robo signals via UNC-34/Enabled and a Netrin-independent UNC-40/DCC function. Nat Neurosci 5(11):1147–1154

Zallen JA, Cohen Y, Hudson AM, Cooley L, Wieschaus E, Schejter ED (2002) SCAR is a primary regulator of Arp2/3-dependent morphological events in *Drosophila*. J Cell Biol 156:689–701

Zheng Y (2001) Dbl family guanine nucleotide exchange factors. Trends Biochem Sci 26:724–732

Zhou Z, Caron E, Hartwieg E, Hall A, Horvitz HR (2001) The *C. Elegans* PH domain protein CED-12 regulates cytoskeletal reorganization via a Rho/Rac GTPase signaling pathway. Dev Cell 1(4):477–489

Zipkin ID, Kindt RM, Kenyon CJ (1997) Role of new Rho family member in cell migration and axon guidance in *C. elegans*. Cell 90:883–894

Chapter 5
Functional Characterization of UNC-53, a Scaffolding Protein During Axon Outgrowth and Cell Migration

Abstract During *Caenorhabditis elegans* development, cells and axons undergo migration along the dorsal ventral and anterior posterior axis to reach their final position to form the final structure and connectivity of the nervous system. While guidance molecules regulating the dorsal ventral guidance have been identified and conserved in both vertebrates and invertebrates, very few guidance molecules have been implicated in anterior posterior guidance and outgrowth. UNC-53, a cytoplasmic scaffolding protein, identified based on uncoordinated phenotype of loss-of-function mutants, regulates growth cone migration along the longitudinal axis of the worm. Interestingly, the vertebrate homologs, neuron navigators (NAVs), also play a role in neurite extension, regeneration, and cell migration. Moreover, expression pattern of UNC-53 and NAVs corroborate with their functional requirement during growth cone migrations. Genetic and biochemical analysis provided insight into the molecular mechanism of UNC-53 action during growth cone migration, suggesting that UNC-53 and NAV may be involved in reorganization of cytoskeleton polymers including actin and microtubules, by physically interacting with molecules like SEM-5, ABI-1, and acting with RhoA stimulator UNC-73E.

Keywords Axon outgowrth · Cell migration · UNC-53 · Neuron navigators

5.1 Introduction

unc-53, a pleotropic gene, was isolated for the first time on the basis of uncoordinated (Unc) phenotype, meaning being incapable of correct reverse locomotion, unlike the wild-type worms, which upon gentle touch on the head move backwards, *unc* mutant worms do not show reverse locomotion (Brenner 1974). Later, several independent studies revealed that mutations in *unc-53* gene result in several behavioral and anatomical defects including defects in excretory canal (EC) cell extension (Hedgecock et al. 1987), a large H-shaped cell, during its outgrowth; two processes emerge from the EC cell body and migrate

A. Pandey and G. K. Pandey, *The UNC-53-mediated Interactome*,
SpringerBriefs in Neuroscience, DOI: 10.1007/978-3-319-07827-4_5,
© The Author(s) 2014

dorsolaterally from the ventral side of the terminal pharyngeal bulb toward the lateral hypodermis, once the canals reach the lateral hypodermis, they cross the hypodermal basement membrane and bifurcate, sending processes anteriorly to the head and posteriorly to the tail (Nelson et al. 1983; Buechner 2002), defects in sex muscles, sex myoblasts (SMs), a set of cells give rise to sex-specific muscles required for normal egg-laying (Sulston and Horvitz 1977; Thomas et al. 1990), at the end of L1 stage two SMs are generated in the left and right ventral muscle quadrants, midway between the center of the gonads and anus, in L2 stage SMs migrate anteriorly and stop precisely when they flank center of the gonads, by L3 stage they undergo three division to give rise to 16 vulval and uterine muscles, *n152* allele was identified based on sex myoblast (SM) defects (Trent et al. 1983) showing Egl phenotype (egg-laying defective). Besides affecting cell migrations, immunocytochemical studies using antiserum for staining neuronal cell bodies and their processes revealed that UNC-53 is required for longitudinal process outgrowth and guidance of the touch neurons. Studies with five different alleles concluded that UNC-53 activity is required for axon outgrowth in the four lateral microtubules (LMs) including ALMs, PLMs, AVM, and PVM (Hekimi and Kershaw 1993), the ALN and PLN neurons, and the motoneurons (Stringham et al. 2002). Besides, *unc-53* mutants are defective in male mating (Hodgkin 1983). Interestingly, overexpression of UNC-53 causes morphological alterations in body muscle cells including exaggerated growth during embryogenesis (Stringham et al. 2002). Structural and mutant analysis confirmed UNC-53 as a cytosolic protein required for GC migration along the longitudinal or AP axis of the worm, amino acid (-aa) sequence analysis revealed that it is composed of multiple domains including actin-binding calponin homology (CH) domain, actin-binding LKKE motif, nucleotide binding site of AAA domain, polyproline-rich region, which binds to SH3 domains, and coiled-coil (CC) region for protein–protein interaction (Stringham et al. 2002).

Homologs of UNC-53 have been identified and studied in other organisms including *Homo sapiens* (*NAV1*, *NAV2*, and *NAV3*) (Maes et al. 2002), mouse (*mNAV1*, *mNAV2*, and *mNAV3*), and zebrafish (*nav3a* and *nav3b*) (Martínez-López et al. 2005; Klien et al. 2011). These were named as Navigators based on the homology with UNC-53 and their role in cell migration and axon path finding in *Caenorhabditis elegans* and mammals.

Retinoic acid inducible in neuroblastoma cells 1 (RAINB1)/NAV2/UNC-53H2/POMFIL2/HELAD1 was first identified in human neuroblastoma SH-SY5Y cell line, a pediatric malignant tumor of neural crest origin (Alexander 2000). RAINB1 distribution is regulated by *all trans* retinoic acid (*at*RA) in SH-SY5Y cells, during embryogenesis *at*RA regulates patterning of the NS, neurite outgrowth, axonal pathfinding, and neuronal regeneration (Gavalas and Krumlauf 2000; McCaffery and Dräger 2000). In vitro studies corroborated the in vivo role of *at*RA, treating SH-SY5Y cells with *at*RA resulted in disaggregation of cells, neurite induction, and inhibition of cell division. In an independent study, RAINB1 was characterized as HELAD1 (helicase, APC-down regulated), mutated in majority of colorectal cancers (Ishiguro et al. 2002). In vivo function of *UNC-53H2/NAV2* studied in hypomorphic mutant mouse, exhibited impairment of sensitivity in various sensory systems including visual, olfactory, pain, and impairment was corroborated by the hypoplasia of the optic nerve (Peeters et al. 2004). In vivo, *NAV2* regulates the normal development of cranial nerves IX (glossopharyngeal) and X (vagus), and in adult mutant mouse, the baroreceptor response requiring the function of these nerves was defective (McNeill et al. 2011). Additionally, analysis of hypomorphic mutant mice lacking full length *NAV2* transcript exhibited defects in cerebellar development due to defects in axon elongation and external granule layer (EGL) neuron migrations (McNeill et al. 2011). Subsequently, *NAV2* was confirmed as an ortholog of *unc-53* by transgene expression of *mec-7::NAV2* in mechanosensory neurons in *unc-53* mutant worms, rescuing the axon extension defects, supporting functional conservation between Navigators and UNC-53 (Muley et al. 2008).

Pore membrane and/or filament interacting like protein 1 (POMFIL1)/NAV3/UNC-53H3, shows homology to rat nuclear pore membrane complex (NPC) protein POM121 (Coy et al. 2002; Hallberg et al. 1993) and UNC-53. POMFIL1 localizes to nuclear membrane and all the three genes exhibit differential NS expression. Its expression is upregulated during brain injury indicative of its role in neuronal regeneration. Based on the reduced expression pattern in primary neuroblastoma and genomic rearrangements in tumor DNA, POMFIL1 is proposed to be a tumor-suppressor gene (Janke et al. 2000; Nikolopoulos et al. 2000). Later studies associated POMFIL1 with mycosis fungoides or Sézary syndrome, the most common form of primary cutaneous T-cell lymphoma (CTCL) using multicolor fluorescent in situ hybridization (FISH) analysis, a translocation was detected, disrupting *NAV3* locus (Karenko et al. 2005).

UNC53H1/NAV1/POMFIL3 is associated with neuronal development and regeneration, and neural tumorigenesis, validated by its increased expression during brain injury (Coy et al. 2002). Studies with mNAV1, mouse navigator, supported the requirement of mNAV1 for guidance during NS development specifically of the pontine cells from the lower rhombic lip (RL) at the leading process. Abolishing mNAV1 activity in in vitro cultures of RL explant cocultured with Netrin-1 secreting cells, resulted in loss of Netrin-1 mediated guidance supporting the requirement of Netrin-1 for mNAV1 guidance function (Martínez-López et al. 2005).

Nav3a, the zebrafish homolog of NAV3 was also shown to be involved in cell migration during development. In vivo time-lapse imaging of liver development in *nav3a* morphants revealed a failure of hepatoblast movement out from the gut endoderm during the liver budding stage (Klein et al. 2011).

Studies undertaken in both vertebrates and invertebrates indicate functional conservation between UNC-53 and neuron navigators (NAVs). In conclusion UNC-53/NAV function is required during embryogenesis for guiding and extension of GC, supported by phenotypic analysis *unc-53/NAV* mutants, exhibiting wide range of nervous-related system disorders related to aberrant growth cone migration.

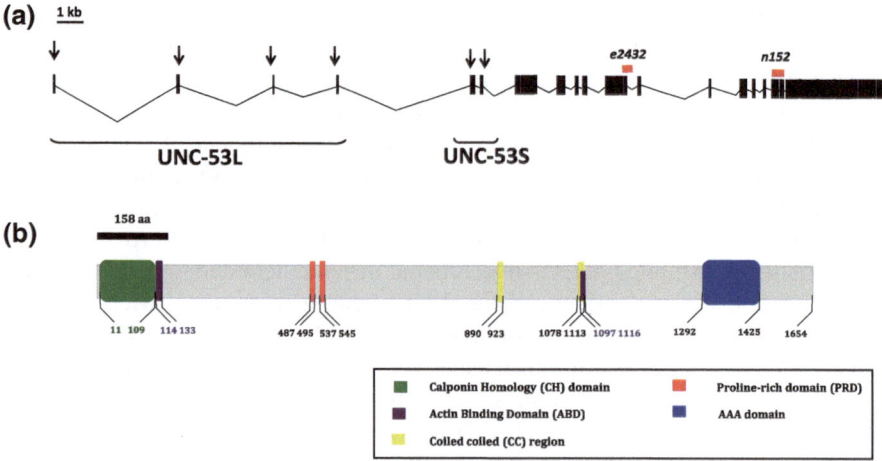

Fig. 5.1 Genomic organization and protein structure of UNC-53: **a** UNC-53 locus is composed of 23 exons (*black blocks*) covering 30-kb genome sequence on linkage group II, has six splice leader (*SL*) sequences marked by *black arrows*, encoding six isoforms categorized into UNC-53L (UNC-53 long) and UNC-53S (UNC-53 short). Various mutations have been mapped onto Unc-53 locus including *e2432* (39 bp deletion, at 28,099–28,462 bp), and *n152* (374 bp deletion, at 2,829–22,227 bp). **b** Unc-53 encodes six isoform, the longest isoform is 1654-aa (amino acid) residue protein, and is composed of N-terminal Calponin homology (*CH*) domain, two proline-rich domains (*PRDs*), actin-binding domains (*ABD*) with LKKE sequence, coiled-coil region (*CC*), and a C-terminus AAA domain. The position of the domains is represented by number of amino acids

5.2 Genomic Organization and Protein Structure of UNC-53 and NAVs

5.2.1 *Genomic Organization and Expression Pattern of UNC-53*

Genetic mapping placed *unc-53* between *bli-1* and *rol-1* locus, on linkage group II (Brenner 1974), covering 31-kb genomic region, composed of 23 exons (Fig. 5.1a) (Stringham et al. 2002). Northern blot analysis identified a 5.0-kb transcript in wild-type worms, and RT-PCR analysis using L2 stage RNA with splice leader (SL1) primer and second antisense primer in exon 9 amplified six transcripts (Pandey and Garriga, unpublished), representing six isoforms categorized into long (UNC-53L) and short (UNC-53S) forms (Fig. 5.1a). All the six transcripts have similar 3′ end but a variable 5′ end, these results corroborated with results published by Stringham et al. (2002).

The expression pattern of *unc-53* was studied using GFP reporter constructs and immunohistochemistry, which correlated with *unc-53* mutant phenotype. Expression was detected in ten head neurons, extending processes in the nerve ring, and ten tail neurons, extending processes anteriorly. Expression was also detected in pioneering neurons detected in comma stage of the embryo, DA motoneurons at threefold embryo stage, and body wall muscles in the embryo stage. In L3 stage larvae, expression was detected in the SMs, subsequent to their anterior migration, and in adult worms at a high level in the vulval muscles vm1 and vm2. In males, expression was seen in the diagonal and spicule retractor muscles (Stringham et al 2002; Schmidt et al 2009).

Later, expression pattern of the long isoform of UNC-53 (UNC-53L) using green fluorescent protein (GFP) and 2.9 kb of genomic region upstream of SL1 site revealed its expression in the head neurons, tail neurons, EC, VNC neurons, and the SMs. Besides, expression was further confirmed in coelomocytes by immunostaining using antibodies raised against the CH domain (or first five exons) (Schmidt et al. 2009). Expression of UNC-53 was observed with *unc-53::GFP* and was coincident with the migration and axon outgrowth of the cells and neurons (Strigham et al. 2002).

5.2.2 UNC-53 Protein Structure

unc-53 transcript encodes for protein products ranging from 1,500 to 1,600-aa residues, the longest product being 1654-aa residues. Starting at the N-terminus, UNC-53 has a CH domain (11–109-aa), approximately 100-aa residue long, required for binding to signal transduction molecules and cytoskeleton binding molecules by directly binding to F-actin including α-actinin and dystrophin, proteins known to cross-link actin filaments into bundles and networks (Hartwig 1995; Stradal et al. 1998; Korenbaum and Rivero 2002). The CH domain is present in UNC-53L but absent in UNC-53S isoform. There are two additional putative actin-binding sites of the LKKE consensus (Van Troys et al. 1999), similar to actin-binding motif of actobindin and villin (Vancompernolle et al. 1992; Stradal et al. 1998), besides it contains two polyproline-rich sequences (PRD), potential SH3-binding motifs (Yu et al. 1994), coiled–coiled (CC) sequences predicted to adopt a CC configuration that could mediate homomeric or heteromeric protein–protein interactions (Maes et al. 2002), at the C-terminus is a nucleotide (NTP)-binding site present within an ATPases associated with diverse cellular activities (AAA) domain, implicated in diverse cellular functions, including cell cycle regulation and vesicle-mediated transport (Fig. 5.1b) (Confalonieri and Duguet 1995).

5.2.3 Genomic Organization, Expression Pattern, and Protein Structure of Neuron Navigators

The three human homologs of UNC-53, the NAVs, were identified in silico. *NAV1* consists of 30 exons and covers a genomic region of 135 kb on chromosome 1q32.1 encoding transcripts ranging from 5.5 to 11.0 kb. Northern analysis supports its expression in fetal brain, similar to mouse navigator *mNAV1*, and in developing NS. *NAV1* expression is down regulated in adult brain, except during pathological conditions,

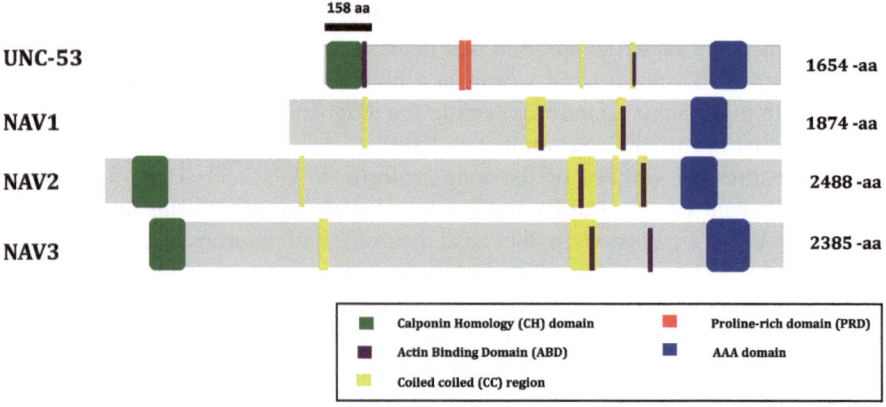

Fig. 5.2 Graphic of protein structure of UNC-53 and NAVs: the domain architecture of UNC-53 and navigator proteins is graphically depicted. Amino acid numbers are given to mark the relative position and length of each domain. The different domains include, AAA domain (*blue*), calponin homology (*CH*) domain (*green*), coiled-coil (*CC*) domain (*yellow*), LKKE actin-binding motif or actin binding domain (ABD) (*purple*), proline-rich domain (*PRD*) (*red*). *C. elegans* has a single gene encoding UNC-53 protein with six isoforms, the longest isoform is represented in the figure coding for 1654-aa residue protein. Homo sapiens genome encodes three Navigators including NAV1, NAV2, and NAV3, which are approximately 1874-aa, 2488-aa, and 2385-aa residue long, respectively. The NAV1 protein does not have the N-terminus CH domain, present in all other NAVs and UNC-53. UNC-53 has two PRDs absent in NAVs

confirmed by SAGE cDNA library of oligodendroglioma, a type of glioma that is believed to originate from the oligodendrocytes of the brain or from a glial precursor cell. The N-terminal part, covering exons 1–8 in *NAV2* and 1–9 in *NAV3*, is entirely missing, coding for the calponin homology (CH) domain, which could alter the functional role of *NAV1* by affecting its localization in the cell. The *NAV1* cDNA encodes around 1874-aa protein. Like UNC-53, it has CC domains (255–280-aa, 1072–1169-aa, and 1303–1362-aa), two LKKE actin-binding motifs (1352–1360-aa and 1148–1155-aa), and a conserved AAA domain (1550–1704-aa) (Fig. 5.2) (Maes et al. 2002).

The complete NAV2 protein corresponding to RAINB1 was separately identified using human neuroblastoma cell line, SH-SY5Y (Merrill et al. 2002). It is present on chromosome 11p15.1 and is spread over 38 exons, translating into approximately 2488-aa protein (Coy et al. 2002). *NAV2* locus encodes transcripts ranging from 7.5 to 9.5 kb, and is expressed in kidneys, heart, and brain. NAV2 is composed of CH domain (90–190-aa), three CC domains (719–752-aa, 1626–1718-aa, and 1841–1896-aa), two LKKE actin-binding motifs (1889–1897-aa and 1702–1708-aa), and AAA domain (1841–1896-aa) (Fig. 5.2) (Maes et al. 2002; Merrill et al. 2002).

NAV3 consists of 39 exons and is located on chromosome 12q21.1. *NAV3* produces transcripts ranging from 7.5 to 9.5 kb, and is expressed in brain. NAV3 is composed of CH domain (1–102-aa), CC domains (596–629-aa, 1476–1570-aa, and 1692–1749-aa), LKKE actin-binding motifs (1554–1561-aa and 1741–1749-aa), and AAA domain (1944–2114-aa) (Fig. 5.2) (Maes et al. 2002).

Structural analysis of UNC-53 and NAVs revealed the absence of PRD domain in NAVs. Presence of other domains in both NAVs and UNC-53 support the conservation at both functional and molecular level.

5.3 The UNC-53 Interactome Involved in Axon Outgrowth and Cell Migration

While the signaling pathways regulating DV guidance are well documented, much less is known about the pathways guiding GCs along AP axis. UNC-53 has emerged as a major protein involved in growth and guidance of GCs migration along the AP axis of the worm. As discussed previously, UNC-53 function is required for both cell migration and axon outgrowth of cell types including SMs, EC, and neurons. Work done in the past several years supports the hypothesis that UNC-53, a multidomain scaffolding protein, interacts with various signaling components to initiate cytoskeletal rearrangements. Some of the keys signaling molecules implicated to work with UNC-53 in longitudinal guidance are discussed in this section.

5.3.1 Cell Migration

unc-53 was identified along with unc-73 and unc-71 in a genetic enhancer screen for mutations affecting SM migration in sem-5 (sex muscle defective) mutants background, SEM-5/GRB-2 (Growth factor receptor-bound protein-2) is an adaptor protein with one Src homology 2 (SH2) and two Src homology 3 (SH3) domains in SH3–SH2–SH3 order, SH2 domain binds to tyrosine phosphorylated proteins. SEM-5 function is required in the gonad-dependent mechanism of SM migration and vulval development (Clark et al. 1992; Chen et al. 1997). Later, biochemical studies confirmed a direct interaction between the PRD of UNC-53 probably through the SH3 domain of GRB2 (Stringham et al. 2002). In SM migration, sem-5 acts downstream of egl-15 encoding a C. elegans fibroblast growth factor receptor (FGFR) (DeVore et al. 1995).

Studies in vertebrates have shown that GRB2 binds to tyrosine phosphorylated C-terminus of EGF (Lowenstein et al. 1992) and downstream of receptor kinase. Similarly, DRK, the fly homolog, acts downstream of Sevenless (receptor tyrosine kinase) and upstream of Sos (son of sevenless), a GEF, and Ras1 (GTPase) in R7 precursor cells (Olivier et al. 1993). In C. elegans, vulval development, SEM-5 acts downstream of Lin-3 (a molecule similar to EGF) and its receptor Let-23 (EGFR homologue), and upstream of Let-60, a Ras (Clark et al. 1992). Based on these studies, it can be hypothesized that probably SEM-5 links UNC-53 to a tyrosine phosphorylated protein, which might be a receptor and maybe these proteins are working in a complex upstream of GTPases during GC movement and guidance.

5.3.2 Axon outgrowth and Excretory Canal Process Extension

UNC-53 has been implicated in extension of EC processes (Stringham et al. 2002). In *unc-53*, mutant background both the processes of EC are severely truncated (Stringham et al. 2002; Schmidt et al. 2009). A number of genes have been implicated in the growth of the EC process including *unc-5*, *unc-34*, *unc-71*, and *unc-73* (Hedgecock et al. 1987). In order to identify interaction between genes involved in EC growth and guidance, RNAi and phenotypic analysis of mutants was done in *unc-53* loss-of-function worms, revealing RhoGEF activity of UNC-73/TRIO (Steven et al. 2005), encoded in the UNC-73E isoform, acting in the same pathway as UNC-53 (Marcus-Gueret et al. 2012). The same study also implicated UNC-73B (RacGEF) and VAB-8 acting in a parallel pathway to UNC-73E and UNC-53. GEF factors stimulate the Rho family GTPases (Lundquist et al. 2001), which in turn affect the actin cytoskeleton. Consistent with this hypothesis, knocking down *rho-1/RHO-A* function enhanced *vab-8* phenotype, supporting the role of UNC-73E isoform in EC process outgrowth and existence of two parallel pathways (Marcus-Gueret et al. 2012). The three different members of Rho GTPases include Rho, Rac, and Cdc-42 (Steven et al. 2005) activate different downstream molecules, where Rho GTPases are known to stimulate formins, which promote long less-branched actin filaments (Evangelista et al. 2003). UNC-71/ADAM has been shown to act along with UNC-53 in the EC process outgrowth. Whether these proteins exist in a complex in EC or not is not known (Marcus-Gueret et al. 2012). Abelson interactor-1 (ABI-1) has been also implicated in EC process and axon outgrowth. Phenotypic analysis of *abi-1* RNAi in *unc-53* mutant background placed them in the same genetic pathway for EC, motoneuron, ALM, and PLM process extension. This function of UNC-53 was associated with the long isoform corroborated by the results of RNAi toward the first four exons. These results were further validated by biochemical studies confirming physical interaction between the first 139-aa of UNC-53 and ABI-1 (Schmidt et al. 2009). ABI-1 is a part of WAVE regulating complex (WRC), and WAVE is regulator of ARP 2/3 complex, responsible for nucleating branched actin filaments, control lamellipodial dynamics (Goley et al. 2006). RNAi in *unc-53* mutant background with *C. elegans* WAVE and WASP complex members, revealed that *wve-1*, *nck-1,* and *arx-2/arp* produced EC migration phenotypes similar to *unc-53* and *abi-1* mutants, while *wsp-1* and *abl-1* did not (Schmidt et al. 2009), supporting the involvement of ABI-1 and components of WAVE complex in actin remodeling via the ARP2/3 complex in UNC-53 pathway. *nck-1* RNAi, an adaptor protein and ABI-1 interactor, exhibited a phenotype similar to *unc-53* suggesting it might play a role in UNC-53-mediated GC movement. MIG-10/Lpd, a MRL adaptor protein family member, acts with UNC-53 in EC process and axon outgrowth, and ABI-1 acts downstream of both UNC-53/NAV and MIG-10/Lpd in this process (McShea et al. 2013). These results point to the existence of more than one pathway involved in UNC-53-dependent and independent GC movement probably both temporally and spatially regulated.

From the above work, it is quite evident that UNC-53 plays a role in actin cytoskeleton remodeling, some recent studies provided further insight into mechanism of UNC-53 action, primarily implicating its function in receptor trafficking. *unc-53* mutant showed phenotypes in coelomocytes uptake (CUP) assay and receptor-mediated endocytosis assays (Fares and Greenwald 2001; Balklava et al. 2007; Stringham and Schmidt 2009). Receptor-mediated endocytosis assay was done in worms with VIT-2::GFP (YP170::GFP) fusion protein, VIT-2, or vitellogenin is the yolk protein, *C. elegans* has five yolk proteins, secreted by intestine into the body cavity, where it is endocytosed by only yolk protein receptor, RME-2, expressed in the oocytes. *unc-53* mutant worms with VIT-2::GFP exhibited defective endocytosis (Pandey and Garriga, "unpublished"), supporting its role in endocytosis. Role of UNC-53 in trafficking was validated by two independent studies including a study in which UNC-53/NAV2 was associated with contactin homolog of *C. elegans*, RIG-6, in GC movement. Contactins are subgroup of immunoglobulin cell adhesion molecules (IgCAM) composed of glycophosphatidylinositol (GPI) membrane anchored glycoproteins with four FNIII-type and six Ig modules, implicated in cellular processes like axon guidance and outgrowth, and neuronal migrations (Rougon and Hobert 2003; Shimoda and Watanabe 2009). Phenotypic analysis of *rig-6* mutants revealed its requirement for mechanosensory, motoneuron, VNC crossing defects, and EC process elongation, phenotypes similar to *unc-53* loss-of-function mutants. Moreover, reduction of *unc-53* function enhanced the RIG-6 overexpression phenotype (Katidou et al. 2013), suggesting that UNC-53 might be involved in RIG-6 endocytosis. Role of UNC-53 in receptor trafficking was supported by studies done in ventral axon guidance of HSN and AVM neurons, where UNC-53 affected the localization of UNC-40 receptor. In wild-type worms with *UNC-40::GFP* transgene, UNC-40 was asymmetrically localized to the ventral side of the HSN neuron in response to UNC-6 guidance cue (Xu et al. 2009). Phenotypic analysis of *unc-53* mutants with *UNC-40::GFP* transgene revealed random UNC-40 localization, resulting in defective axon outgrowth, suggesting UNC-53 function in proper localization of UNC-40 receptor (Kulkarni et al. 2013). It can be concluded that UNC-53 steers GCs along the longitudinal axis probably by affecting trafficking of receptors, utilizing actin cytoskeleton for this purpose.

5.3.3 Navigators in Nervous System Development

Navigators function in GC guidance and movement during embryogenesis in MT-dependent manner, supported by studies done with mouse *mNAV1* (Neuron navigator 1), expressed in the developing NS and implicated in signaling downstream of Netrin-1/UNC-6 in directional guidance in pontine-migrating cells. In vitro studies further showed that *EGFP::mNAV1* transfected cells resulted in microtubule (MT) bundling, and EGFP::mNAV1 binding to MT, this association was mediated by a N-terminal noncanonical microtubule binding domain (MTBD) (Martínez-López et al. 2005). Besides, EGFP::mNAV was detected in filopodial GC. Another evidence in support of role of MT came from the experiments done with

NAV2/RAINB1, inducing neurite outgrowth in SH-SY5Y cells in response to *at*RA. Transfection of *NAV2/unc-53* constructs in Cos-1 cells revealed a region of the protein (837–1065-aa) that directs localization with the MT cytoskeleton, a localization disrupted upon MT destabilization (Muley et al. 2008). Moreover, NAV2 localization both in the cell body and along the length of the growing neurites of SH-SY5Y cells closely mimicked that of neurofilament and MT proteins. Collectively, these works support a role for NAV2 in neurite outgrowth and axonal elongation by facilitating interactions between MTs and other proteins such as neurofilaments that are key players in the formation and stability of growing neurites (Muley et al. 2008). Additionally, it was shown that all the navigators were plus end tracking proteins (+TIPS), +TIPs specifically associate with the ends of growing MTs, and are involved in many cellular processes, including mitosis, cell migration, and neurite extension. Overexpression of GFP-tagged Navigators displaced CAP_GLY-motif containing +TIPs, such as CLIP-170, from MT ends, suggesting that the Navigator-binding sites on microtubule ends overlap with those of the CAP_GLY-motif proteins. Fluorescence recovery after photobleaching (FRAP) experiments in interphase cells, indicated that NAV1 localizes to the centrosome and associates with intracellular structures other than MTs. Expression of GFP-tagged Navigators induces the formation of neurite-like extensions in non-neuronal cells, showing that Navigators dominantly alter cytoskeletal behavior, requiring ATPase activity for this function. Later, using a live imaging approach, it was shown that all three Navigators are capable of tracking MT ends in Cos7 cells. Furthermore, in vivo localization studies with antibodies against the Navigators revealed that the endogenous proteins also accumulate at MT ends (van Haren et al. 2009).

Using the cytoskeletal interacting region of NAV2/UNC-53 in yeast two-hybrid, identified a novel NAV2 interactor, 14-3-3ε, a regulatory protein which binds to diverse ligands including transmembrane receptors, signaling proteins, phosphatases, and kinases. The *Drosophila* homolog D14-3-3ε function is required for pole cell migration during embryogenesis (Tsigkari et al. 2012). RNAi studies in *C. elegans* using *ftt-2*, worm homolog of 14-3-3ε, resulted in defects similar to *unc-53* RNAi (Marzinke et al. 2013). 14-3-3ε proteins are conserved across the species (Aitken et al. 1992; Aitken 2006; Skoulakis and Davis 1998), and function in diverse cellular processes including signal transduction, cell cycle regulation, apoptosis, and protein localization. Moreover, they also participate in neuronal migration (Toyo-oka et al. 2003).

5.4   Mechanism of UNC-53/NAV Action in GC Migration and Guidance

Functional characterization of UNC-53 and its human homologs, the NAVs revealed that these proteins promote GC migration of cells and axons during embryogenesis, primarily acting in the nervous system, evidenced by genetic and biochemical studies in *C. elegans* and mammals (Stringham and Schmidt 2009;

McNeill et al. 2011). In vitro, experiments with the mouse homolog, mNAV1, showed NAVs functions at the leading edge of filopodia GC to promote neurite induction in response to Netrin-1/UNC-6 (Martínez-López et al. 2005). The functional characterization was further validated by molecular characterization including expression analysis in *C. elegans* and mammals, confirming expression of UNC-53 and NAV transcripts primarily in nervous system (Stringham et al. 2000; Maes et al. 2002; Schmidt et al. 2009). These results were further corroborated by phenotypic analysis of *unc-53* mutants exhibiting pleiotropic phenotype including defects in nervous system development.

Studies in past several years have pointed that UNC-53 and NAVs promote GC extension by directly interacting with signaling molecules, through their various domains, including adaptor protein like SEM-5/GRB2, through the PRD (Stringham et al. 2000), cytoskeleton polymers including actin via the CH domain (Stringham et al. 2000) and MTs through the noncarnonical MTBD (van Haren et al. 2009), ATP, and GTP molecules via the Walker motif of AAA domain (van Haren et al. 2009), and ARP 2/3 regulating protein, ABI-1 through the CH domain (Schmidt et al. 2009). Functional and molecular characterization of UNC-53 suggests that it affects GC extension by regulating actin cytoskeleton reorganization. Two important assays, which revealed the mechanism of UNC-53 action included the CUP (coelomocyte uptake) assay and receptor-mediated endocytosis assay in *C. elegans* coelomocytes and oocytes respectively, UNC-53 exhibits positive CUP assay and defective receptor uptake (Fares and Greenwald 2001; Balklava et al. 2007; Stringham et al. 2009; Pandey and Garriga, unpublished), thereby providing a link between UNC-53-mediated cytoskeleton rearrangement and endocytosis. Moreover, actin and MTs have been implicated in endocytosis, where MTs regulate the late steps of endocytosis, disrupting MTs inhibited translocation of endosomes and lysosomes (Matteoni and Kris 1987; Gruenbarg et al. 1989; Gottlieb et al. 1993). Interestingly, both UNC-53 and NAVs have the actin binding (CH domain) and a noncanonical MTBD, it can be speculated that UNC-53/NAV function by binding actin and MT to regulate endocytosis (Strigham et al. 2002; van Haren et al. 2009). Studies in *C. elegans* further support the role of UNC-53 in actin remodeling, as UNC-53 directly binds to WRC component ABI-1, a regulator of ARP 2/3 (Schmidt et al. 2009). Moreover, *unc-53* mutation affected the localization of UNC-40::GFP in HSN and AVM neurons and also enhanced RIG-6 overexpression phenotype (Katidou et al. 2013; Kulkarni et al. 2013). All these results validate UNC-53 requirement in receptor trafficking.

A molecular model can be proposed for UNC-53 interactome involved in regulating endocytosis. Given its direct physical interaction with SEM-5/GRB2, an adaptor protein known for interacting with tyrosine phosphorylated receptor; UNC-53 might be indirectly linked to a receptor. These two molecules physically interact with the PRD of UNC-53 and probably SH3 motif of SEM-5, this complex might further stimulate UNC-73E, GEF molecule. Evidenced by studies in *C. elegans*, *Drosophila*, and mammals where GRB2 functions downstream of receptors and upstream of Ras proteins (Clark et al. 1992; Lowenstein et al. 1992; Olivier et al. 1992). Similarly, in *C. elegans* UNC-53 functions with UNC-73E in

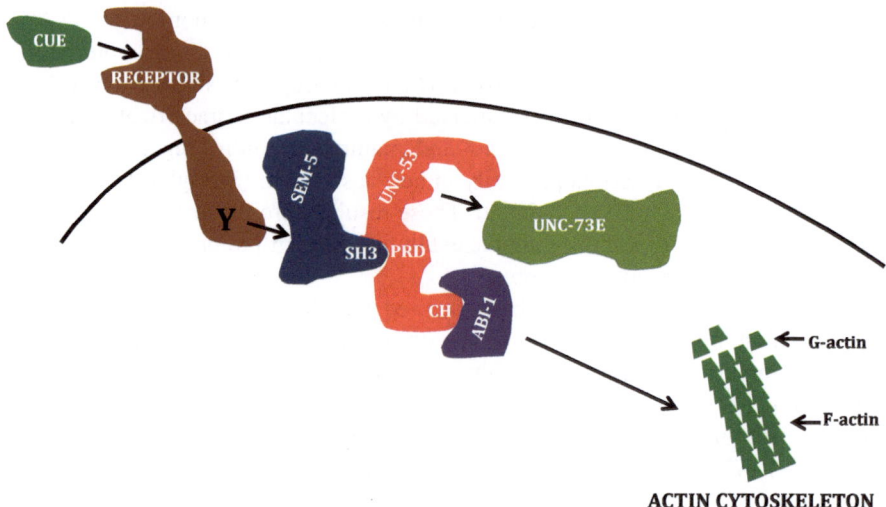

Fig. 5.3 Molecular model for UNC-53 interactome regulating actin cytoskeleton remodeling: UNC-53 is required for growth cone migrations along the longitudinal axis of the worm. Genetic and biochemical analysis has supported the role of SEM-5, ABI-1, and UNC-7E to function with UNC-53 in regulating the outgrowth and guidance. Binding assays have confirmed that UNC-53, which is a cytosolic protein, interacts with SEM-5/GRB2, an adaptor protein. This interaction is between the PRD region of UNC-53 and SH3 region of SEM-5 and ABI-1 via its CH domain. Analogous to vertebrates and fly homolog GRB-2, SEM-5 might form a complex at the membrane containing UNC-53, a tyrosinated receptor, and ABI-1 in response to external cue. ABI-1 is a component of WAVE regulatory complex (*WRC*), which in turn regulates ARP2/3

EC and mechanosensory axon guidance to activate downstream RhoA (Marcus-Gueret et al. 2012), an actin cytoskeleton modulator. Alternatively, UNC-53 physically interacts with ABI-1 to regulate actin nucleator ARP 2/3 complex (Fig. 5.3). The interaction of UNC-53 with different molecules to generate an interactome or signaling module may vary spatially and temporally to affect GC movement.

References

Aitken A (2006) Seminars Cancer Biol. 16:162–172

Aitken A, Collinge DB, van Heusden BP, Isobe T, Roseboom PH, Rosenfeld G, Soll J (1992) Trends Biochem Sci 17:498–501

Alexander F (2000) Neuroblastoma. Urol Clin North Am (3):383–392, vii

Balklava Z, Pant S, Fares H, Grant BD (2007) Genome-wide analysis identifies a general requirement for polarity proteins in endocytic traffic. Nat Cell Biol 9:1066–1073

Brenner S (1974) The genetics of *Caenorhabditis elegans*. Genetics 77:71–94

Buechner M (2002) Tubes and the single *C. elegans* excretory cell. Trends Cell Biol 12:479–484

Chen EB, Branda CS, Stern MJ (1997) Genetic enhancers of *sem-5* define components of the gonad-independent guidance mechanism controlling sex myoblast migration in *Caenorhabditis elegans* hermaphrodites. Dev Biol 182(1):88–100

Clark SG, Stern MJ, Horvitz HR (1992) *C. elegans* cell signaling gene *sem-5* encodes a protein with SH2 and SH3 domains. Nature 356:340–344

Confalonieri F, Duguet M (1995) A 200-amino acid ATPase module in search of a basic function. BioEssays 17:639–650

Coy JF, Wiemanna S, Bechmannb I, Bächnerc D, Nitschb R, Kretzd O, Christiansene H, Poustka A (2002) Pore membrane and/or filament interacting like protein 1 (POMFIL1) is predominantly expressed in the nervous system and encodes different protein isoforms. Genes 290:73–94

DeVore DL, Horvitz HR, Stern MJ (1995) An FGF receptor signaling pathway is required for the normal cell migrations of the sex myoblasts in *C. elegans* hermaphrodites. Cell 83:611–620

Evangelista M, Zigmond S, Boone C (2003) Formins: signaling effectors for assembly and polarization of actin filaments. J Cell Sci. 116: 2603–2611

Fares H, Greenwald I (2001) Genetic analysis of endocytosis in *Caenorhabditis elegans*: coelomocyte uptake defective mutants. Genetics 159:133–145

Gavalas A, Krumlauf R (2000) Retinoid signalling and hindbrain patterning. Curr Opin Genet Dev 10:380–386

Goley ED, Ohkawa T, Mancuso J, Woodruff JB, D'Alessio JA, Cande WZ, Volkman LE, Welch MD (2006) Dynamic nuclear actin assembly by Arp2/3 complex and a baculovirus WASP-like protein. Science. 314:464–467

Gottlieb T, Ivanov I, Adesnik M, Sabatini D (1993) Actin microfilaments play a critical role in endocytosis at the apical but not the basolateral surface of polarized epithelial cells. J Cell Biol 120:695–710

Gruenbarg J, Howell KE (1989) Membrane traffic in endocytosis: insights from cell-free assays. Annu. Rev. Cell Biol. 5:453–481

Hallberg E, Wozniak RW, Blobel G (1993) An integral membrane protein of the pore membrane domain of the nuclear envelope contains a nucleoporin-like region. J Cell Biol 122:513–521

Hartwig JH (1995) Actin-binding proteins. 1: Spectrin super family. Protein Prof 2(7):703–800

Hedgecock EM, Culotti JG, Hall DH, Stern BD (1987) Genetics of cell and axon migrations in *Caenorhabditis elegans*. Development 100:365–382

Hekimi S, Kershaw D (1993) Axonal Guidance defects in a *Caenorhabditis elegans* mutant reveal cell-extrinsic determinants of neuronal morphology. J Neurosci 13(10):4254–4271

Hodgkin J (1983) Male phenotypes and mating efficiency in *Caenorhabditis elegans*. Genetics 103:43–64

Ishiguro H, Shimokawa T, Tsunoda T, Tanaka T, Fujii Y, Nakamura Y, Furukawa Y (2002) Isolation of HELAD1, a novel human helicase gene up-regulated in colorectal carcinomas. Oncogene 21(41):6387–6394

Janke J, Schluter K, Jandrig B, Theile M, Kolble K, Arnold W, Grinstein E, Schwartz A, Estevez-Schwarz L, Schlag PM, Jock-usch BM, Scherneck S (2000) Suppression of tumorigenicity in breast cancer cells by the microfilament protein profilin 1. J Exp Med 191:1675–1686

Karenko L, Hahtola S, Päivinen S, Karhu R, Syrjä S, Kähkönen M, Nedoszytko B, Kytölä S, Zhou Y, Blazevic V, Pesonen M, Nevala H, Nupponen N, Sihto H, Krebs I, Poustka A, Roszkiewicz J, Saksela K, Peterson P, Visakorpi T, Ranki A (2005) Primary cutaneous T-cell lymphomas show a deletion or translocation affecting NAV3, the human UNC-53 homologue. Cancer Res. 65(18):8101–8110

Katidou M, Tavernarakis N, Karagogeos D (2013) The contactin RIG-6 mediates neuronal and non-neuronal cell migration in Caenorhabditis elegans. Dev Biol. 373(1):184–195

Klein C, Mikutta J, Krueger J, Scholz K, Brinkmann J, Liu D, Veerkamp J, Siegel D, Abdelilah-Seyfried S, le Noble F (2011) Neuron navigator 3a regulates liver organogenesis during zebrafish embryogenesis. Development 138(10):1935–1945

Korenbaum E, Rivero F (2002) Calponin homology domains at a glance. J Cell Sci 115:3543–3545

Kulkarni G, Li H, Wadsworth WG (2008) CLEC-38, A Transmembrane Protein with C-Type Lectin-Like Domains, Negatively Regulates UNC-40-Mediated Axon Outgrowth and Promotes Presynaptic Development in Caenorhabditis elegans. J Neurosci. 28:4541–4550

Lowenstein EJ, Daly RJ, Batzer AG, Li W, Margolis B, Lammers R, Ullrich A, Skolnik EY, Bar-Sagi
    D, Schlessinger J (1992) The SH2 and SH3 domain-containing protein GRB2 links receptor
    tyrosine kinases to ras signaling. Cell 70(3):431–442
Lundquist EA, Reddien PW, Hartwieg E, Horvitz HR, Bargmann CI (2001) Three *C. elegans*
    Rac proteins and several alternative Rac regulators control axon guidance, cell migration and
    apoptotic cell phagocytosis. Development 128:4475–4488
Maes T, Barcelo A, Buesa C (2002) Neuron navigator: a human gene family with homology to
    unc-53, a cell guidance gene from *Caenorhabditis elegans*. Genomics 80:21–30
Marcus-Gueret N, Schmidt KL, Stringham EG (2012) Distinct cell guidance pathways controlled
    by the Rac and Rho GEF domains of UNC-73/TRIO in *Caenorhabditis elegans*. Genetics
    190:129–142
Matteoni R, Kreis TE (1987) Translocation and clustering of endosomes and lysosomes depends
    of microtubules. J Cell Biol 105:1253–1265
Martínez-López MJ, Alcántara S, Mascaró C, Perez-Branguli F, Ruiz-Lozano P, Maes T, Soriano
    E, Buesa C (2005) Mouse neuron navigator 1, a novel microtubule-associated protein
    involved in neuronal migration. Mol Cell Neurosci 28(4):599–612
Marzinke MA, Mavencamp T, Duratinsky J, Clagett-Dame M (2013) 14-3-3e and NAV2 interact
    to regulate neurite outgrowth and axon elongation. Arch Biochem Biophys 540:94–100
McCaffery P, Dräger UC (2000) Regulation of retinoic acid signaling in the embryonic nervous
    system: a master differentiation factor. Cytokine Growth Factor Rev 11(3):233–249
McNeill EM, Klöckner-Bormann M, Roesler EC, Talton LE, Moechars D, Clagett-Dame M
    (2011) *Nav2* hypomorphic mutant mice are ataxic and exhibit abnormalities in cerebellar
    development. Dev Biol 353(2):331–343
McShea MA, Schmidt KL, Dubuke ML, Baldiga CE, Sullender ME, Reis AL, Zhang S, O'Toole
    SM, Jeffers MC, Warden RM, Kenney AH, Gosselin J, Kuhlwein M, Hashmi SK, Stringham
    EG, Ryder EF (2013) Abelson interactor-1 (ABI-1) interacts with MRL adaptor protein
    MIG-10 and is required in guided cell migrations and process outgrowth in *C. elegans*. Dev
    Biol. 373(1):1–13
Merrill RA, Plum LA, Kaiser ME, Clagett-Dame M (2002) A mammalian homolog of *unc-53* is
    regulated by all-trans retinoic acid in neuroblastoma cells and embryos. Proc Natl Acad Sci
    USA 99(6):3422–3427
Muley PD, McNeill EM, Marzinke MA, Knobel KM, Barr MM, Clagett-Dame M (2008) The
    atRA-responsive gene neuron navigator 2 functions in neurite outgrowth and axonal elongation.
    Dev Neurobiol 68:1441–1453
Nelson FK, Albert PS, Riddle DL (1983) Fine structure of the *Caenorhabditis elegans* secretory-
    excretory system. J Ultrastruct Res 82:156–171
Nikolopoulos SN, Spengler BA, Kisselbach K, Evans AE, Biedler JL, Ross RA (2000) The
    human non-muscle alpha-actinin protein encoded by the ACTN4 gene suppresses tumori-
    genicity of human neuroblastoma cells. Oncogene 19:380–386
Olivier JP, Raabe T, Henkemeyer M, Dickson B, Mbamalu G, Margolis B, Schlessinger J, Hafen E,
    Pawson T (1993) A *Drosophila* SH2-SH3 adaptor protein implicated in coupling the sevenless
    tyrosine kinase to an activator of Ras guanine nucleotide exchange, Sos. Cell 73(1):179–191
Hafen E, Pawson T (1993) A Drosophila SH2-SH3 adaptor protein implicated in coupling the
    sevenless tyrosine kinase to an activator of Ras guanine nucleotide exchange, Sos. Cell
    73(1):179–1191
Peeters PJ, Baker A, Goris I, Daneels G, Verhasselt P, Luyten WH, Geysen JJ, Kass SU,
    Moechars DW (2004) Sensory deficits in mice hypomorphic for a mammalian homologue of
    unc-53. Brain Res Dev Brain Res. 150(2):89–101
Rougon G, Hobert O (2003) New insights into the diversity and function of neuronal immuno-
    globulin superfamily molecules. Annu Rev Neurosci 26:207–238
Schmidt KL, Marcus-Gueret N, Adeleye A, Webber J, Baillie D, Stringham EG (2009) The cell
    migration molecule UNC-53/NAV2 is linked to the ARP2/3 complex by ABI-1. Development
    136:563–574

Shimoda Y, Watanabe K (2009) Contactins: emerging key roles in the development and function of the nervous system. Cell Adhes Migr 3:64–70

Skoulakis EM, Davis RL (1998) Mol Neurobiol 16:269–284

Steven R, Zhang L, Culotti JG, Pawson T (2005) The UNC-73/Trio RhoGEF-2 domain is required in separate isoforms for the regulation of pharynx pumping and normal neurotransmission in *C. elegans*. Genes Dev 19:2016–2029

Stradal T, Kranewitter W, Winder SJ, Gimona M (1998) CH domains revisited FEBS Lett 431:134–137

Stringham E, Pujol N, Vanderchove J, Bogaert T (2002) *unc-53* controls longitudinal migration in *C. elegans*. Development 129:3367–3379

Stringham E, Schmidt KL (2009) Navigating the cell UNC-53 and the navigators, a family of cytoskeletal regulators with multiple roles in cell migration, outgrowth and trafficking. Cell Adhes Migr 3(4):342–346

Sulston JE, Horvitz HR (1977) Post-embryonic cell lineages of the nematode *Caenorhabditis elegans*. Dev Biol 56:110–156

Thomas JH, Stern MJ, Horvitz HR (1990) Cell interactions coordinate the development of the *C. elegans* egg-laying. Cell 62:1041–1052

Toyo-oka K, Shionoya A, Gambello MJ, Cardoso C, Leventer R, Ward HL, Ayala R, Tsai LH, Dobyns W, Ledbetter D, Hirotsune S, Wynshaw-Boris A (2003) Nat Genet 34:274–285

Trent C, Tsung N, Horvitz HR (1983) Egg laying defective mutants of the nematode *Caenorhabaditis elegans*. Genetics 104:619–647

Tsigkari KK, Acevedo SF, Skoulakis EM (2012) 14-3-3ε Is required for germ cell migration in Drosophila. PLoS One 7(5):e36702

Vancompernolle K, Goethals M, Huet C, Louvard D, Vandekerckhove J (1992) G- to F-actin modulation by a single amino acid substitution in the actin binding site of actobindin and thymosin beta 4. EMBO J 11:4739–4746

van Haren J, Draegestein K, Keijzer N, Abrahams JP, Grosveld F, Peeters PJ, Moechars D, Galjart N (2009) Mammalian Navigators are microtubule plus-end tracking proteins that can reorganize the cytoskeleton to induce neurite-like extensions. Cell Motil Cytoskelet 66(10):824–838

Van Troys M, Vandekerckhove J, Ampe C (1999) Structural modules in actin-binding proteins: towards a new classification. Biochim Biophys Acta 1448:323–348

Xu Z, Li H, Wadsworth WG (2009) The roles of multiple UNC-40 (DCC) receptor-mediated signals in determining neuronal asymmetry induced by the UNC-6 (netrin) ligand. Genetics 183:941–949

Yu H, Chen JK, Feng S, Dalgarno DC, Brauer AW, Schreiber SL (1994). Structural basis for the binding of proline-rich peptides to SH3 domains. Cell 76:933–945